HUNTING DINOSAURS

HUNTING DINOSAURS

LOUIE PSIHOYOS
WITH JOHN KNOEBBER

RANDOM HOUSE
NEW YORK

TITLE PAGE: Dinosaur model maker Peter May of Research Casting International, Toronto, Canada, puts the final touches on an Allosaurus cast bound for the front hall of the American Museum of Natural History in New York.

PAGE V: Bam Bam on Dino racing against Mongolian boy

PAGE VII: A researcher at the Mongolian State Museum in Ulan Bator assembles the remains of an ankylosaur discovered in the Gobi Desert. The heavy body armor and massive tail clubs of these creatures made them a formidable opponent for even the largest of predators.

PAGE 1: The Flaming Cliffs of the Gobi Desert, made famous by Roy Chapman Andrews's expedition of the 1920s.

 Published in the United States by Random House, Inc., New York, and simultaneously in Canada by Random House of Canada Limited, Toronto.

All photographs, unless otherwise indicated, were taken by Louie Psihoyos.

Library of Congress Cataloging-in-Publication Data

Psihoyos, Louie.
Hunting dinosaurs / Louie Psihoyos with John Knoebber.
p. cm.
Includes bibliographical references.
ISBN 0-679-43124-1
1. Paleontology. I. Knoebber, John. II. Title.
QE711.2P78 1994 567.9'1—dc20 94-10010

Manufactured in the United States of America

98765432

First Edition

DESIGNED BY JOEL AVIROM AND JASON SNYDER

To Vicki, Nico, and Sam

— L.P., 1994

———

To those who love travel, adventure, food, and life

— J.K., 1994

Being a paleontologist is like being a coroner except all the witnesses are dead and all the evidence has been left out in the rain for 65 million years.

—Michael Brett-Surman,
Smithsonian Institution, Washington, D.C.

FOREWORD

DINOSAURS—
NATURE'S SPECIAL EFFECTS

L ouie Psihoyos's book is a splendid mix of scientific photojournalism and firsthand adventure. It's the story of men and women who go out on a very special type of big-game hunt, a hunt that takes them across vast stretches of geological time, to bring back spectacular creatures from the realm of the Deep Past.

OPPOSITE: Paul Sereno's expedition cruises through the canyons of the Los Colorados Formation near the Valley of the Moon, Argentina, in search of the earliest dinosaur.

It was in the fourth grade that I got hopelessly hooked on dinosaurs. What grabbed my attention was one piece of scientific journalism—a long, beautifully written story in *Life* from September 7, 1953, with *Brontosaurus* on the cover and, inside, the whole saga of evolution through the ages. Picutres by *Life* staff photographers, prehistoric scenes created by painter Rudy Zalinger, and the text by Lincoln Barnett went together to make a powerful story.

As I read the article, I decided that I would dedicate my life to Mesozoic paleontology—I'd help find missing pieces to the dinosaur family saga.

Several of my present colleagues tell similar tales about that 1953 *Life* issue; they owe their career choice to this single article. A second, very powerful influence on us budding dinosaurophiles of the 1950s was Roy Chapman Andrews's *All About Dinosaurs*, still one of the best kids' books on the topic. Andrews's account was unapologetically first-person. He was the expedition leader of the great American Museum forays into the Gobi Desert in the 1920s and 1930s. Andrews wasn't a dinosaurologist himself. But his prose captured the extraordinary moment when the first dinosaur eggs were unearthed in the fire-red sandstones of the Flaming Cliffs.

Andrews wrote for *National Geographic* too—one of my prized possessions is a June 1933 issue with an essay on the Gobi, complete with tinted photographs of the *Protoceratops* nests and eggs, as well as views of the *Baluchitherium* quarry, which yielded the bones of an extinct, long-legged rhino that was as big as a *Diplodocus*.

Sometimes we Ph.D. types get a bit stuffy and cliquish. Some-times we use so much scientific jargon that we obscure the pure joy of dino hunting. When we write "scholarly" monographs about the braincases of *Ankylosauria*, we can get mired in descriptions of the basicranial foramina for the *ramus ophthalmicus, nervi trigeminalis* (sensory nerves that go to the tip of the snout), and the osseous walls of the *fossa orbitalis* (eye socket). And sometimes we forget how important are clear, first-person narratives that capture the excitement of bone digging.

That's why all us dinosaurophiles, both amateur and professional, need Louie Psihoyos's book.

Louie is in the tradition of the classic photojournalists who made *National Geographic* famous. He has tramped through dozens of bug-infested badlands. He's pushed his camera lens into a score of quarries cut into rocks from the Triassic, Jurassic, and Cretaceous periods. His photos catch the drama of bone hunters walking and crawling down countless arroyos, getting cactus thorns in their hands and silt in their beards, hoping to find that next missing link. Louie has recorded dinosaurs in the act of being freed from their sedimentary tombs, as patient fingers pry away the rock one small chip at a time.

And Louie has followed the fossils into the laboratory and the exhibit hall, where the petrified skeletons rise on hidden metal armatures. Fossils are tough subjects for photoportraiture. The petrified bones have uneven surfaces and carry an unpredictable mosaic of textures and mineral stains. I've seen Louie take twelve hours to set up lights and reflectors to get just one shot of a *T. rex*.

The results are stunning. Skulls and rib cages, I know well as a scientist, take on a new status as natural sculptures of extraordinary grace. There has always been a strong element of aesthetics in paleontology—we bone diggers endure poor financing and bad bug bites because the bones we find really are beautiful. Louie's photographs affirm that belief.

Louie told me he was surprised at himself when he sat down to write the text. He already knew that he could shoot film—his colleagues and competitors rank him as the number one paleophotojournalist. But he didn't realize that he could write too. He can.

It's very rare to find someone with a keen visual sense who also can narrate his own work. Some Ph.D. types offered to coauthor Louie's book, figuring that we could help with our vast background knowledge, but he wisely turned us down. Louie has the ability you see in veteran Associated Press reporters: he can join a bunch of paleontological specialists, all talking technospeak, and ask a few questions that get to the center of the issue. And he has an ear for the punchy quote from a professor that sums up a lifelong philosophy of digging and interpreting bones.

Louie has produced the best, most vivid account of bone sleuths, past and present, since Chapman and Barnett a half century ago. And Louie's book goes much further than any other has. He covers the most recent, breaking news—the discovery of missing links that help complete the dino family history. A striking example is Paul Sereno's discovery in Argentina of *Eoraptor*, a beast very close to the basal trunk of the dino family tree. Louie covers forgotten tragedies, such as the sinking of an English freighter with a near-perfect mummy of a duckbill dinosaur in the cargo hold by the German secret raider *Moewe* in the First World War. And he covers the dino bloopers and farces, most notably the seizure by the F.B.I. of a *Tyrannosaurus* skeleton that was on public display, freely accessible to everyone. (The government, of course, claimed it had to seal up that skeleton in crates in order to protect it for the People of the United States. This reminds me of the last scene in *Raiders of the Lost Ark*, where a government forklift unloads the Ark in the bowels of a nameless federal warehouse.)

Louie carries around with him an expensive hardwood box with a brass clasp. Inside, it is lined with felt, like the carrying case of a Fabergé Egg or some other jewel-encrusted ornament of the Czar. If Louie thinks you have the proper level of respect for the science of bone sleuthing, he may open the box and show you what's inside.

It's a man's skull, with a bulging braincase and a delicate face. The owner of this cranium was Professor Edward Drinker Cope of Philadelphia. Cope and his rival, Professor Marsh of Yale, fought the Bone Wars in the 1870s and 1880s, racing each other to find all the great dinosaurs preserved in the rocks of Colorado, Wyoming, and the Dakota Territory. Cope, by all accounts, was the better scientist, with an unusually broad knowledge of living salamander, frog, and fish species, as well as their fossil kin.

Cope was a paleontological swashbuckler. Professor Marsh and his Ivy League colleagues set out on expeditions with a support column of dozens of students and quarrymen, convoyed, on occasion, by troops of cavalry. Cope probed the Missouri Breaks for dinosaurs in the 1870s accompanied only by a Kansas farm boy, disregarding dangers of Indian war parties and terrible drought. And Cope's quick wit charmed nearly everyone he met—even Marsh's foreman at Como Bluff had a hard time not liking him.

Cope could enter a drawing room in Philadelphia or a saloon in Montana or a ranch in Wyoming and make friends with people from both high and low stations in society. There's a lot of Cope in Louie Psihoyos.

Interest in dinosaurs is not a fad. Dinosaurs are nature's special effects, extraordinary monsters that were created not by a Hollywood computer-animation shop but by the natural forces of evolution. The dinosaur story is a ripping yarn that spans nearly 200 million years and includes the origin of our own, warm-blooded mammalian clan. Dino science always has benefited from a large dose of volunteerism—people donate weekends and summer vacations to digging and cleaning and analyzing bones.

The best way to recruit dino hunters, full-time and part-time, and the best way to generate support for the science is through outstanding pieces of paleontological journalism. Louie Psihoyos's book is the best so far.

— Bob Bakker

ACKNOWLEDGMENTS

First we'd like to thank all the folks at *National Geographic*, who have pursued excellence with a vengeance and jetted us through the looking glass with a fistful of credit cards despite the screams and wails of the accountants. There are too many key supporters in this noble enterprise to mention, but a few are: editor Bill Graves, and Bill Allen, for giving us enough rope to potentially hang ourselves; director of photography Tom Kennedy and his right-hand man, Kent Kobersteen; former directors of photography and guiding lights of photojournalism Bob Gilka and Rich Clarkson; former editor Bill Garrett; picture editors Mary and Tom Smith and Science Editor Rick Gore; layout editors David Griffin and Bill Marr.

We'd like to give special thanks to Illustrations Editor Bill Douthitt, the irreverent genius savant whose insanity rivals anyone's with a good job. If you want to take pictures for the magazine, or just share your darkest private thoughts, Bill is always willing to talk, especially late at night. He's listed in the phone book.

Also at *National Geographic*, we'd like to thank Charlene Valeri, Thelma Altemus, Lilian Davidson, and Andrea Strudwick, who have been tireless supporters throughout; in the camera department ("The *National Geographic* Toy Store"), Nelson Brown, Larry Kinney, and Sergio Ballivian; in the travel department, Ann Chenoweth, Ann Judge, Lillian Giacone, and Mike Mitchell. And a hearty thanks to all the support staff, the unsung heroes who give *Geographic* its soul, and whom, I suspect, really run the place.

To Stella, my mom and our dinosaur accountant, who stuck by us long after the money ran out, love always. Thanks to our brothers, Gus Psihoyos and Roger Knoebber, for general brotherliness and logistic and moral support throughout. Also to Louie's brother-by-law, Ronny Bromberg, for letting us use his Pennsylvania farm to conduct research, and to Ruth and Howard Bromberg for their love and support throughout. And special thanks to the whole Knoebber clan.

And of course to all the paleontologists, curators, and preparators, past and present, for whose love of nature and quest to understand its wisdom and share its beauty we give our deepest heartfelt thanks.

Special thanks to Bob Bakker for his inspired and irreverent counsel, to his lovely and brilliant wife, Constance, and their canine, Prance.

Thanks also to Kevin Padian of the University of California, Berkeley, for his encouragement, advice, and technical guidance.

In Italy, to our hyperlinguistic divine guide, Father Guiseppe Leonardi, dinosaur tracker extraodinaire who is worth several volumes alone!

In Argentina, to Luis Chiappe, Guillermo Rougier, Fernando Novas, and Paul Sereno and the Ischigualasto crew, Andrea Arcucci, Claudia Marsicano, Catherine Forster, Cathleen May, and Romer prize winner Ray Rogers, Oscar Alcober, "Wa Wa Wa" (nobody really needs to know what that means), and our driver Santiago "Duck" Nielsen, who safely navigated us through the Valley of the Moon; in Tucumán, to Jamie Powell; in Plaza Huincul, to Rodolfo Coria and Pedro A. Leonardi. In Salta, to Bustos 1 and Bustos 2 (Ricardo Alonso) and the three Amigos, Pablo Dib Ashur, Ben Heit, and Esteban Tálano. *Muchas gracias mi amigos de dos estúpidos norte Americanos.* In Patagonia, to Marcelo, the rancher who broke bread with us and shared his wine and humble dirt-floor abode.

In Canada, to Phil Currie, for opening his house and sharing his expertise, fine home-brewed beer, and wretched rice wine. To Dave, our climbing instructor with the endless supply of Daffy Duck T-shirts, who saved my life one sunny day on the face of a cliff. To Hans Larsson, Wendy Sloboda, and Bill Sarjeant for opening his impeccable files for us. To Linda Strong-Watson, Brian Noble, P. J., and all the Ex Terra folks. To Peter May of Research Casting International. To Andrew Leitch in Toronto for his CAT-scanning skills. To artist Michael William Skrepnick for his rendering of "Sue." To Gerhard Maier for sharing his knowledge of Tendagura dinosaurs.

In England, to Angela Milner at the British Museum of Natural History for her help in early dinosaur history.

In Washington, D.C., to Bob "Mr. Africa" Caputo for his moral support and embassy. To Dale Newbury at the National Bureau of Standards and Thomas Hardt at Electro Scan for their SEM assistance. To Declan Haun and the Odyssey group for their early support.

In Los Angeles, to Steven Spielberg and Marvin Levy of Amblin Entertainment and all the people at Stan Winston Studio who worked on *Jurassic Park,* including our contributing *Jurassic Park* artists who worked on this book, Shannon Shea and Michael Trcic.

In New Mexico, to Wilson and Peggy Bechtel, who are preparing *Seismosaurus.* To Roland Hagan and his crew at the Los Alamos National Laboratories. To sculptor Dave Thomas.

In Arizona, to Vince Santucci, paleontologist, and his crew at Petrified Forest National Park.

In Russia, to our multitalented interpreter and duck wrangler, Ludmila Mekertycheva.

In Connecticut, to Jack McIntosh, who gave freely of his expert knowledge and extensive files.

In Gotham, to Perry Rubenstein, art wizard, who gave his support and the keys to his Manhattan palace. To my agents, Barbara Sadick and John Wells of Matrix International. To Gotham stylist Tracy Garett. At the American Museum, to Mark Norell, Michael J. Novacek, Phil Fraley, Elizabeth Chapman, and Joel Sweimler.

In Wyoming, to Brent Breithaupt, curator at the Geological Museum at the University of Wyoming, for his Como Bluff stories.

In Paris, to Philippe Taquet and Armand de Ricqlès for their relentless support. And to our French floozies, Helen Guetary and Mary Sloan.

In Bristol, to Peter Crowherd; in Berlin, to Wolf Dieter Heinrich; in Stuttgart, to Rupert Wild; in Munich, to Peter Wellnhofer; and in Caen, France, to Michel Rioult, for their help in tracking down dinosaurs lost in World War II.

To West Coasters Chuck and Carla Levdar for wiring me into the computer age and providing West Coast office support. To artist and psychedelic warrior Larry Fuente, who taught that being weird isn't good enough but will do in a pinch. To our legal literary shark, Brad Bunnin of Berkeley, and Karen Pearson, who skillfully lined the box of our traveling companion, Edward Drinker Cope.

In Colorado, to Boulderites Tom and Wendy Kahn and contributing artist Pat Redman, the Salvador Dali of the Mesozoic. To Mark and Pam Arbini for lending us their home for research. In Denver, to Martin Lockley and Tom Kelly of the Colorado Dinosaur Tracker Society and transcribers Barbara Wehrfritz and Gerry Fourney.

In Utah, to Wade Miller, director of the Earth Science Museum at Brigham Young University, and his compadre Ken Stadtman.

In Mongolia, to the prime minister for the use of his helicopter. To Altangerel Perle for opening the Mongolian collections to us and escorting us to the Flaming Cliffs. To Rinchen Barsbold.

To Shelley Bowen for convincing me to write. This book is mostly her fault.

In the Black Hills, all the best to Pete and Neal Larson, Bob Farrar, Terry Wentz, and Sue the tyrannosaur, of the Black Hills Institute. To Wally from the Alpine Inn, a great place to stay if you are ever in Hill City, South Dakota.

To all those coprolite collectors who gave us crap, including Bill Branson.

To Mike Caldwell, who in two hours told me more about the history of life than I had learned in a lifetime.

On the Navajo Reservation, to Nolan, Shann, Jason, Floyd Stevens, and the Navajo Tribal Council.

To our editor-in-exile, Carolyn White.

In the afterlife, to our spiritual guide, Professor Edward Drinker Cope, with whom we shared a wonderful, long, strange trip.

And most important, to Executive Editor Dave Rosenthal and Associate Publisher Walter Weintz of Random House, who believed in this project from the beginning.

To those of you we forgot or neglected to mention, we apologize. We owe you a beer. Put your name here:

To _____.

May the Winds and Gods be with you always.

—Louie Psihoyos and John Knoebber

CONTENTS

INTRODUCTION

THE GREAT DINOSAUR TOUR GOES INTO DEEP TIME; OR, WHAT A LONG, STRANGE TRIP IT'S BEEN

Dinosaurs were the last thing on my mind. I was sitting in five thousand square feet of prime New York City real estate, a luxury steel-and-concrete bunker with wraparound views of the glitter and grime of Gotham. Before I moved into this Manhattan loft, the landlord bragged that from the windows you could witness a crime an hour. Ten years ago New York had my kind of wildlife.

I was several years into a career photographing the better-known *Homo sapiens* when Bill Douthitt called. Bill is an old friend from my days at *National Geographic*, where I have worked on and off for the last fifteen years. He calls now and then to polish his warped sense of humor and try to lure me back for the odd job.

To be an assignment photographer for Bill has become what I imagine it's like to be a very good plumber to royalty: I am invited in only when there is a particularly nasty problem story, usually one severe enough so that anybody else with a normal cerebral cortex wouldn't touch it. I could almost hear the water rising around Bill's neck.

A woman wrote a letter to Gene Gaffney, Curator of Fossil Reptiles at the American Museum of Natural History in New York, who thought there were too many dinosaurs represented in the displays and suggested some of them be ground into fertilizer, her idea being that the bones were high in mineral content and therefore they might be useful for something. Gaffney wrote her back and admitted that they would probably make good fertilizer, but to him, dinosaurs are objects of awe and beauty and grinding them up for fertilizer would be akin to grinding up Michelangelo's *David* because it would make pretty good cement mix.

—*Paul Sereno, August 1992*

OPPOSITE: Two fine *Photo sapien* specimens, John and I, are admired by grade-school visitors at the National Museum of Natural History in Paris, where modern paleontology began in the eighteenth century with Baron Georges Cuvier.
Photo by Shelley Bowen

"How would you like to do a story for us? It involves a lot of foreign travel."

"What's the story?" I asked cautiously. One of the first stories Bill came up with for me was "The Fascinating World of Trash," an epic journey of discovery through the world's notable garbage dumps. (This became the cover story of *National Geographic*, April 1983.) The whole thing started out as a joke in the lunchroom, as we were inventing articles we would like to see in the relentlessly optimistic magazine: "Our Friend, the Maggot: Life Goes On Inside a Corpse" and "Bulldozer Across America" were examples. Then Bill saw someone taking out the trash and we began brainstorming a shooting list for a garbage story. After our laughter died, we realized that there was probably a real story there. I was a summer intern then; come fall I would be jobless. I wrote up a proposal, it was accepted, and I became the first contract photographer to be hired by the magazine in eleven years. My first year was spent entirely in garbage dumps around the world.

"Dinosaurs," he told me.

ABOVE: Dinosaur stamp from Antigua, where the author lived between expeditions and where the real estate is 36 million years too young to have been a nursery for dinosaurs.

BELOW: John, with club, and Louie outside a rock shop near Petrified Forest National Park in Arizona.

"Dinosaurs!" I said incredulously. "You mean photographing old dead things? Thanks, pal," I told Bill before cutting the conversation short. I was busy running up a phone bill getting back to people who would be too busy to get back to me. But he called again and again over the next month, leaving cryptic messages on my answering machine and slowly chipping away at my resistance to turn my talents to old dead things.

"You're perfect for this. You have Caravaggio's eye and the mind of Gary Larson," he said in one message as an attempt at flattery. "Aren't you sick of that meaningless New York celebrity tripe?"

Meaningless New York celebrity tripe is a very good living, I thought as the ubiquitous sirens screamed down the street below. But I had to admit that I needed a change. Famous people and their edgy handlers were growing old for me. "Dinosaurs," Bill spoke darkly, "are more famous than any human. They will sit still for a photograph as long as you like and they don't need publicists."

One great thing about working for the little yellow magazine is that you can live anywhere. I moved my family from Manhattan to an island of similar size, Antigua in the West Indies. Out my front door I had a bay roughly the size of New York Harbor but with no other houses in view. Well, one other house. The water was clear; I could watch pelicans diving for supper, a pair of giant leopard rays jumping, and dolphins herding sprat. I learned to dive for lobster and spear fish for my supper. I put my shoes in the closet, and there they stayed until I hopped on a plane for some distant dinosaur locality with my best friend, John.

I met John Knoebber on the trash story. He was living in the redwoods of Mendocino helping an artist make surrealistic pieces from garbage that they called "found objects." Some of their "trash" is at the Smithsonian in Washington, and one piece became the cover of the April 1983 issue of *Geographic.* John also became an invaluable friend and assistant, and over the last fifteen years we have traveled around the world on several stories. He's an incredible organizer, a dedicated worker, and has the uncanny ability to take a dull subject and milk it for more fun than seems possible. He is one of those rare people who have no limits. His foot is still on the fantasy accelerator long after I would have slammed on the brakes. As a kid living in Ft. Lauderdale, Florida, in the early sixties, John turned his backyard pool into an aquarium. The first resident was an eight-foot (244 cm) mako shark. At age fifteen he ran away from home and jumped aboard the old wooden schooner *Western Union* on her last working voyage to mend the telegraph cable linking the United States and Cuba. Later his bid for a formal education ended when he was rejected by the only college to which he ever applied, the Ringling Bros. and Barnum & Bailey Clown College. He has been an urchin diver, a leather worker, a building contractor, an artist and sailor, and has had several other colorful careers best left unlisted and for which he has never been convicted. For me, he is what Huck Finn was to Tom Sawyer. I refused to do the dinosaur story unless John could assist me. Insanity was a serious ingredient for the miracle it would take to bring dinosaurs back to life.

Dinosaurs were around for 165 million years. Man, depending on where you personally feel comfortable drawing the line between yourself and another ape, has been here for between 3 million years (*Australopithecus*) and 90,000 years (*Homo sapiens*). Geologically speaking, we are newcomers to the planet. Dinosaurs were the biggest, the most diverse, and (if you accept the current evidence that birds are the liv-

ing descendants of dinosaurs) the longest surviving animals ever to inhabit the land.

There are more people in dinosaur research than ever, each reading different chapters of the earth's crust, like a magnificent journal whose epic stories of fight and survival lie scattered in the sediments around the world. Scientists are chipping away at the evidence, piecing back together different parts of this incredible story.

At the end of our magazine coverage, we realized we might have more than a cover story as dinosaurs became our obsession. Between pictures and interviews, we started looking at the ground for signs of ancient life. We have made modest contributions to the field, and several of our finds are in museum collections around the world. We discovered dinosaur eggs in Patagonia and the Gobi Desert, dinosaur footprints in Canada and Utah, and dinosaur bones all over the world. I had been keeping a journal of our adventures, and later we went back around the world and recorded conversations with scientists.

With forty-two cases of lighting, camera gear, and recording equipment, we traveled over 250,000 miles (402,250 k), shot over 2,000 rolls of film, and assembled what we believe is the most comprehensive dinosaur expedition ever.

—Louie Psihoyos,
Antigua, November 12, 1992

HUNTING
DINOSAURS

IN THE BEGINNING...

THERE WAS NO RENT

There are some dispassionate seers of science that view the miraculous parade of life from the creation of the first cell to the present bouquet of life-forms as a random act, like the kind of dumb luck that wins the lottery. But Nature is a more calculating winner. Chance has played a part in life from the very beginning, from the first cell replicating itself, but after that, success was programmed. From the DNA level, life was encoded to adapt to an ever-changing environment. Life is finely honed, and by its very nature, destined to succeed.

Dinosaurs, in their day, were the epitomes of success, and some say they still are—they took to the air and live on through their descendants, the birds. What we know about them and all creatures before us comes largely through paleontologists, those tour guides of the past who combine aspects of the arts and all the sciences and blend them into a vision that illuminates the darkness of Deep Time.

I have met presidents and princes, exiled royalty, and celebrities of every ilk, but never have I been so thrilled as when making the acquaintance of the paleontologists in this book and the creatures they have brought back to life.

Bon voyage, my friends!

And out of the ground the Lord God formed every beast of the field, and every fowl of the air; and brought them unto Adam to see what he would call them: and whatsoever Adam called every living creature that was the name thereof.

—Genesis 2:19, Authorized (King James) Version

The direct lineage of human ancestry is a hotly debated topic, but the following progression sketches possible sister groups in the fossil record that may have led to man's ascension.

1. The first cell replicates itself about 3.6 billion years ago.

2. Cyanobacteria, also called blue-green algae, at 3.465 billion years ago, is one of the first known complex microorganisms.

3. *Pikaia* is the first known chordate relative, at 420 million years ago. Chordates include all the vertebrates, including humans.

4. Thelodont: About 400 million years ago, jawless fishes develop.

5. *Eusthenopteron,* at 350 million years ago. Lungfish possessed the lobefins that may have enabled it to pioneer the land.

6. *Ichthyostega,* at 350 million years ago. This tetrapod may have been the transitional form between lungfish and amphibians.

7. *Dimetrodon,* at 270 million years ago. Early mammal-like reptiles were the first top predators on land.

8. *Thrinaxodon,* at 250 million years ago. The best known of the earliest mammals.

9. *Morganucodon,* at 200 million years ago. True mammals shrunk to rat-sized critters coming out at night, while dinosaurs ruled the day.

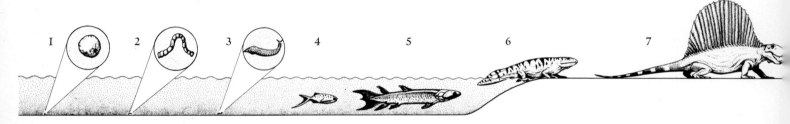

10. *Asioryctes,* at 75 million years ago. Early placental mammal.

11. Plesiadapiformes (*Saxonella*), at 60 million years ago. An early primate. The fossil record for man, chimps, and apes from 4 to 14 million years ago is essentially empty.

12. *Cercopithecus,* at 4 million years ago. This Old World monkey is still around today.

13. *Australopithecus afarensis,* at 3.2 million years ago, is commonly thought of as the first hominid. It walked upright, had an opposable thumb, and possibly used tools.

14. *Homo erectus,* 1 million to 700,000 years ago. Considered our direct ancestor.

15. *Homo sapiens,* 90,000 years ago to the present.

Illustration by Dana Geraths
with revisions by Pat Redman

DINOSAUR PREHISTORY

A briefing on the history of time is necessary for any expedition into the unimaginable expanse of the earth's history that geologists call Deep Time. By using radiometric dates of meteors, the best scientific minds of our day place the age of the earliest known rocks, and hence the age of the earth, at about 4.6 billion years. A staggering number to us taxpaying mortals who live as best we can one day at a time. Measuring the earth's history against our own paltry life expectancy of seventy years is as incomprehensible as balancing the national deficit out of our personal checkbook.

Kevin Padian of the University of California at Berkeley uses the following analogy to give his students a Deep Time perspective. "If you take the history of life as the length of your arm, then one stroke of a nail file erases human history."

Putting it another way, geology professor Don L. Eicher* came up with this brilliant analogy of compressing the earth's history into one calendar year. With some of my own updated and twisted additions we find that on

January 1:	The earth begins.
Springtime, March 20:	The birthday of DNA. The first one-celled bacteria, bobbing happily in the muck, re-creates itself. All life-forms thereafter will be stamped with this same DNA.
Thanksgiving:	Sea creatures begin pioneering the land.
December 11:	90 percent of all life-forms go extinct.
December 13:	Dinosaurs enter.
The day after Christmas:	Dinosaurs go extinct.
The evening of December 31:	Manlike creatures appear.
December 31, 11:59:45 to 11:59:50:	Roman Empire rises and falls.
3.5 seconds to midnight:	Columbus discovers America (or, if you wish, Indians discover Columbus).
1/20th of a second to midnight:	The Beatles play the *Ed Sullivan Show*.

* Don L. Eicher, *Geologic Time* (Englewood Cliffs, N.J.: Prentice Hall, 1976).

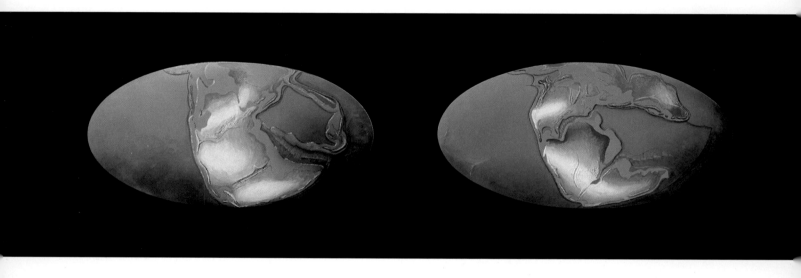

EARLY TRIASSIC
(237 million years ago)

It becomes apparent why dinosaurs are found world-wide. During the Triassic Period (248–208 million years ago), when dinosaurs began to evolve, all the earth's real estate was joined into one supercontinent called Pangaea (All Earth). At the end of the Triassic Period the continents and their emerging dinosaur fauna began sailing apart at the estimated rate of one centimeter (0.39 in) per year—about the rate a fingernail grows. A fast-moving plate travels about 2½ centimeters (0.98 in) per year.

LATE JURASSIC
(152 million years ago)

During the Jurassic Period (208–144 million years ago)—over millions of years, through billions of earthquakes, large and small—mountains were created (including the Jura Mountains of France and Switzerland, which give the Jurassic Period its name), and oceans formed, isolating plant and animal faunas and royally messing up the weather.

THE BREAKUP
OF THE EARTH

*Map illustrations by C. R. Scotese,
University of Texas at Arlington*

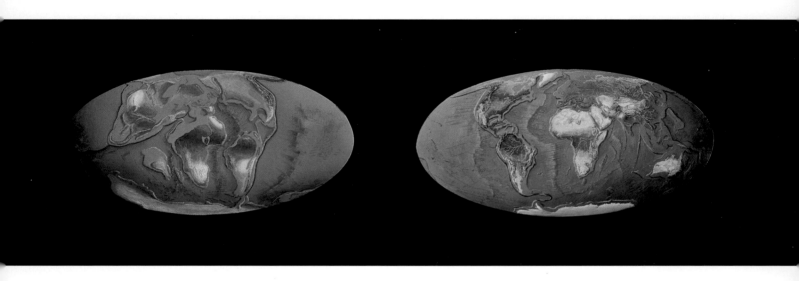

LATEST CRETACEOUS
(65 million years ago)

THE EARTH TODAY

Using fossilized plants and animals is one of the more useful ways to calculate where the pieces of Pangaea once fit together. For instance, coal miners of Appalachia and England are extracting coal created by the same geologic formation.

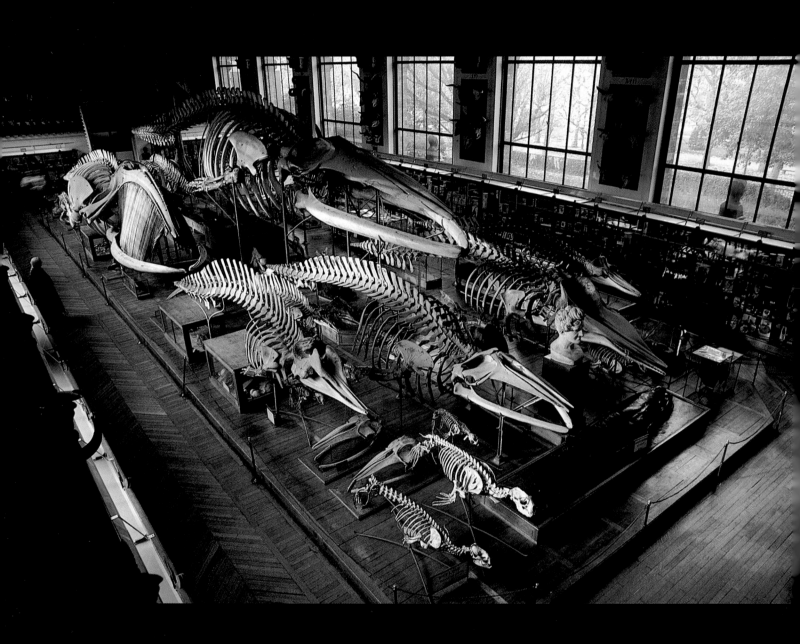

THE EARLY BONE HUNTERS

BARON GEORGES CUVIER (1769–1832), the French scientist, is considered to be the father of modern paleontology and comparative anatomy. During Cuvier's time it was a widely held conception that animals were created by an all-perfect God, and it was inconceivable that He would allow His creations to become extinct. It was Cuvier who popularized the idea of extinction and debunked the myth that all creatures still existed somewhere on unexplored parts of the planet.

OPPOSITE: The National Museum of Natural History in Paris was the workshop of Baron Georges Cuvier. Around 1800, Cuvier was describing the bones of strange, extinct creatures unearthed from the Montmartre gypsum quarries, the original source of plaster of paris. His creations stunned Europe and awed his fellow Parisians. Exclaimed Honoré de Balzac, "Is Cuvier not the greatest poet of our century? Our immortal naturalist has reconstructed worlds from blanched bones. He picks up a piece of gypsum and says to us 'See!' Suddenly stone turns into animals, the dead come to life, and another world unrolls before our eyes." *

*From *The Riddle of the Dinosaur* by John Noble Wilford (New York: Random House, 1987).

GIDEON MANTELL (1790–1852), a Sussex country doctor and enthusiastic fossil collector, was the first person to recognize the remains of a giant reptile. One day in 1822, while he made a house call, his wife, Mary Ann, found the teeth of a beast he would later call *Iguanodon*, after the iguana, whose teeth it resembled. His contemporaries trivialized the find, and it wasn't until 1825 that he had the courage of his convictions to write up a scientific description. In 1844 Mantell, an eloquent writer, authored a book about fossils called the *Medals of Creation* in which he explained fossils as medallions struck by the Creator to commemorate the success of epochs.

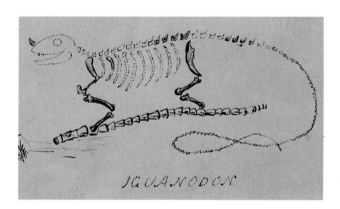

TOP LEFT: Gideon Mantell
Department of Library Services,
American Museum of Natural History, New York
TOP RIGHT: Mary Ann Mantell
The Natural History Museum, London
LEFT: The first specimens of *Iguanodon* (iguana tooth) were teeth discovered by Mary Ann Mantell from the lower Cretaceous (about 110 million years ago) stones of the Bestede Quarry, Cuckfield, Sussex. They are now in the collections of the Natural History Museum in London.
ABOVE: Working eventually with more *Iguanodon* remains, Mantell made the first reconstruction of a dinosaur.
The Natural History Museum, London

WILLIAM BUCKLAND (1784–1856),
a theologian and professor at Oxford University, kept strange animals about the house and church grounds and brought his pet bear, Tiglath Pileser (say it ten times fast), to wine parties dressed in cap and gown. Besides being fun to party with and having his umbrella engraved "Stolen from William Buckland," this legendary eccentric is also noted for publishing the first scientific description of a dinosaur, *Megalosaurus*, in 1824.*

*Fun bits distilled from *The Great Dinosaur Hunters and Their Discoveries* by Edwin H. Colbert (New York: Dover Publications, 1968, 1984).

SIR RICHARD OWEN (1804–1892),
Her Majesty Queen Victoria's knighted paleontologist—royal hobnobber and Darwin-basher and, by most frank accounts, not a very likable guy but a brilliant scientist—was the man responsible for giving dinosaurs their name. Before the British Association for the Advancement of Science in 1841, Owen proposed the name "Dinosauria," from the Greek which means, roughly, "fearfully great lizard," to describe the newly discovered tribe of creatures being hauled out of European landscapes. Although not all dinosaurs were fearful (some recently found dinosaurs were chicken-sized at best) and none were lizards, the name has stuck. Working with only a few fragmentary remains, Owen envisioned dinosaurs as scaled up lizards. Despite this handicap, in 1854 Owen oversaw construction of a dinosaur park at the Crystal Palace outside London that inspired the first Dinomania on record. Hundreds of thousands of visitors flooded the park to get a glimpse of the incredible life-size models (bottom right) of dinosaurs.

CHARLES DARWIN (1809–1882) never encountered dinosaurs on his various odysseys, but he did gather data for his revolutionary treatise on evolution, *On the Origin of Species*. Darwin explained life as a parade of mutations evolving through a process he called natural selection. The foundations of Victorian England were shaken by Darwin's contradiction of the biblical version of creation.

THOMAS HENRY HUXLEY (1825–1895) was left by Darwin, never one to argue in public for his own controversial ideas, to champion his friend and colleague's evolutionary theory. Called "Darwin's bulldog," Huxley fearlessly argued evolution with Sir Richard Owen and the queen's bishops as the messy task of setting the record straight began.

For fifty years after Gideon Mantell first discovered extinct reptiles, European dinosaur researchers had only scrappy remains—until a mother lode of fully articulated dinosaurs came to light in 1878. Coal miners working in a shaft 1,046 feet below the surface of Bernissart, Belgium, thought they had discovered a tree trunk full of gold. What they really found were the remains of a herd of plant-eating *Iguanodons.* The *Iguanodon* herd suffered from pyrite disease, an affliction that causes fossilized bone, when subjected to moisture, to suddenly crack and dissolve into a fine bright yellow powder—pyrite, sometimes called fool's gold. These specimens, pictured with curator P. Bultynck at the Royal Institute of Natural Sciences in Brussels, considered a national treasure, are now stabilized and kept in a glass enclosure to help reduce the humidity.

What's in the Box?

OR,
The Strange Travels of Edward Drinker Cope—
Specimen 4989

One of the greatest bone hunters who ever lived, Professor Edward Drinker Cope, was our traveling companion for the three years we conducted our field research on dinosaurs. Our association with this great professor came as quite a surprise, as he had been dead for nearly a hundred years.

Until his death in 1897, Cope, a consummate naturalist, had named over twelve hundred new species of animals. By the age of ten he was already drawing detailed anatomical studies of lizards, and at the age of twenty he had published dozens of papers in scientific journals. In spite of a pacifist upbringing by his Quaker father, Edward had cultivated a fiery temperament and a passion for brawling. When the Civil War began, Cope had sympathies for the Union, but to prevent him from joining the army, his father sent him off to study in Europe. Ironically, while exiled in Europe, Cope would meet the man with whom he would wage the most bitter battle in modern science.

THE GREAT BONE WARS

Othniel Charles Marsh was a late bloomer and liked to be called O. C. He entered Yale at twenty-five, so old that his fellow students called him "Daddy." Marsh, who initially aspired to be a machinist, decided to build a career in science to get the financial support of his rich uncle George Peabody, a brilliant banker and London-based financier who was known to open his billfold wide to enterprising researchers. Marsh, never a brilliant student but a canny manipulator, talked his uncle into bankrolling $150,000 to start a museum at Yale, provided the institution could secure Marsh the directorship, a post he held throughout his Yale career. Although Marsh never married, he was a devoted social climber who entertained lavishly in his eighteen-room mansion near the college. He ventured out west on fossil-collecting missions for only three seasons (1870–1872), but the stories he collected from his brief sojourns entertained parlor guests for the rest of his life. He used his new social position as head of the Peabody Museum to amass power and employ field-workers and researchers whom he seldom acknowledged as we would today. All the famous animals he is credited with naming (*Brontosaurus, Stegosaurus, Triceratops*) were discovered by his researchers.

Although Cope was also born into wealth and employed crews, he remained a field person his whole life, continuing his search even in the midst of the great Indian Wars and encouraging and acknowledging his field help.

Like most sworn and bitter enemies, Cope and Marsh started out as friends. After the Civil War they collected fossils in the field together in the East. Both were already powerful figures in the field, Marsh director of the Yale College Museum and Cope a famous paleontologist at the Academy of Natural Sciences in Philadelphia. At this early stage of their friendship, each had named new species after the other.

As legend has it, the feud began when Marsh visited Cope's dino fields in New Jersey and met his collectors. Marsh surreptitiously paid these men to send future fossils to him. That was the start.

Then in 1870 Marsh pointed out, quite correctly, that the strange new *Elasmosaurus*, a giant long-necked, long-tailed sea reptile that Cope had described in a noted scientific journal, was only unique because of

ABOVE: Specimen 4989
OPPOSITE: Professor Edward Drinker Cope (1840–1897),
one of the most accomplished bone hunters in history,
described over 1,200 vertebrates and published over
1,400 scientific papers.

ABOVE: Portrait of O. C. Marsh
BELOW: Cope's home in Philadelphia (A) and Marsh's home in New Haven (B).

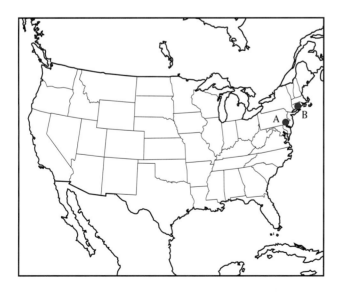

the way Cope had reconstructed it—he had put the beast's head at the wrong end. The embarrassment to Cope was shattering and left a permanent scar on his significant ego. Cope began a campaign to buy up all the copies of the published journal, but Marsh kept his as future testimony to his colleague's ineptitude. A feud began that raged their whole adult lives.

In the annals of modern science there never was a rivalry so bitter or so productive. Paleontologists, in twisted reverence, call this period the Great Bone Rush, or the Great Bone Wars, because the last twenty years of Marsh and Cope's embattled lives were the most productive in dinosaur history until the present boom in discoveries. In each man's quest to find more animals than the other, more than 130 of the 287 species of dinosaurs then known were described by Cope and Marsh.* Their rival field crews became masters of stealth and deception and as paranoid as miners protecting gold claims. They and their teams would often travel under pseudonyms, digging up bones at new sites as quickly as possible, then smashing the remaining ones to render the site useless to their rivals. Both sides employed spies to rustle dinosaurs from the other's unsuspecting camp. A species of the famous predator *Allosaurus*, credited to Cope, was rustled from Marsh by one of Cope's enterprising field crews at Como Bluff, Wyoming. Field teams would ship the bones back to their employer, who would race to publish a description first. In their haste and greed to outdo each other, they gave many of the same animals several names. A slight variation in size might warrant a new name. In the case of one *Uintatheres anceps*, a mammal vaguely resembling an elephant with no trunk but having six blunt horns on its head, the two bone hunters, paleontologist Bob Bakker told me, collectively granted twenty-two names.

The Bone Wars were made public when the *New York Herald*, in the heyday of yellow journalism, let each side use its front pages as a battlefield. For several days each scientist launched slanderous volleys accusing the other of impropriety of funds, libel, and worse. Though litigation appeared inevitable and lawyers were mobilized, neither side could ultimately claim victory: Cope

*There are now (October 1993) 949 known species of dinosaurs. A new species is found about every six weeks.

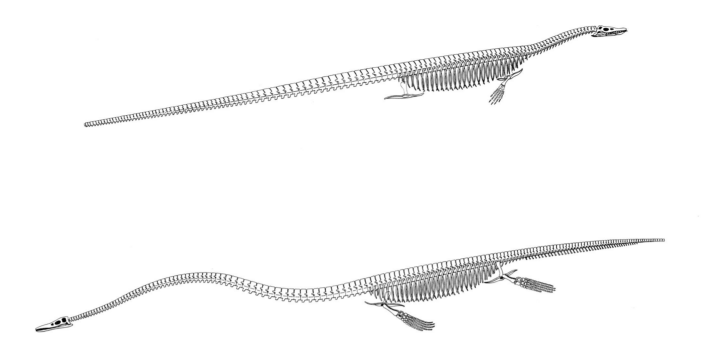

had to sell his coveted collection of fossils to pay his mounting bills, and Marsh, who launched many of his expeditions on the coattails of U.S. geological survey teams, saw the bulk of his fossils confiscated by the federal government for the Smithsonian Institution, the national repository of natural history. In the end it was the museums that triumphed—the myriad dinosaurs that Cope and Marsh collected filled their great halls.

BONE WARS VETERAN EDWARD DRINKER COPE BECOMES OUR SPIRITUAL GUIDE

I wanted to illustrate the Great Bone Wars with an assemblage of Cope and Marsh artifacts. To bring together this unlikely still life, it was necessary to conduct a raid on the curatorial vaults of six East Coast museums and then assemble all of the objects in one location. Museums, however, do not easily part with their sacred relics. By a stroke of luck we discovered some famous Marsh drawings that had been hidden away in the basement of the Smithsonian by a paleontologist who had rescued them from his curator's trash. It was a spectacular find, and our interest in them validated the historical significance the paleontologist had seen in them. He happily lent us the rare drawings. When curators of other museums found out that we had obtained the Marsh drawings, they tried, uncon-

ABOVE: Legend has it that the Great Bone Wars escalated in a major way when Marsh, on a visit to Cope in Philadelphia, pointed out that the strange new *Elasmosaurus*, a marine reptile that Cope assembled, was only unique in the way he had assembled it—its head at the wrong end. The embarrassment to Cope was immense. He began a campaign to buy up all the offending copies he had published in the *Transactions of the American Philosophical Society* and had a new engraving made and paid for a republication. Here, published together, are the famous versions; the top reconstruction is Professor Cope's suppressed version, which we found in the rare-book collection of the Princeton Archives.

OVERLEAF: Artifacts from the lives of archenemies O. C. Marsh (left) and Edward Drinker Cope. From Yale University, the Marsh pick became the standard for today's paleontologists. Marsh's commissioned drawings of a *Ceratosaurus*, from the archives of the Smithsonian Institution, provide a backdrop for his compass and portrait of him (center row middle) and his 1870 field crew to the West. Cope artifacts include: his pick and field diary from the American Museum of Natural History; from the Smithsonian archives, headlines of the original *New York Herald* chronicling their public feud; field specimens discovered in the vaults of the Philadelphia Academy of Natural Sciences, left as Cope had prepared them for shipment—still wrapped in newspapers of the day, the *Fargo Forum* and the *Sioux County Herald*, both dated 1893. From the University of Pennsylvania, the bones of the legendary bone hunter himself, Professor Edward Drinker Cope.

sciously perhaps, to top this Smithsonian loan. By the time we arrived at the Academy of Natural Sciences in Philadelphia, where Cope was once curator, John and I had acquired a serious historical archive.

Professor Cope's modern counterpart at the academy, Ted Daeschler, was reconstructing a mental list of the Cope artifacts he could make available when he mentioned that Cope himself was on a museum shelf across town. Professor Cope, Daeschler related, had a rather unusual death wish. Scientists of the past centuries, in their rush to discover new animals, had overlooked collecting the most obvious animal of all—modern man. For those scientists who discover and name new animals, taxonomists, the most coveted trophy is called the type specimen—the first described animal of a new species. That specimen becomes the measure by which all other members of that species are judged. By actually becoming the type specimen for man, *Homo sapiens,* Cope arguably would become, taxonomically anyway, the ultimate man and have an ultimate last laugh over his nemesis Professor Marsh, who would die a few years later.

Per his instructions, Cope was prepared for his postmortem career as *Homo sapiens* type specimen and his bones were dedicated to science. But science, as it turned out, didn't want them. His bones were badly decalcifying—showing, it appeared, the beginning signs of syphilis. They were deemed unworthy as the type specimen and shelved at the Wistar Institute in Philadelphia, a repository for anatomy specimens. Years later, when Wistar shifted its emphasis to biomedicine, it gave much of its fossil and anatomy collections to the University of Pennsylvania, which then inherited the esteemed Professor Edward Drinker Cope. He was put in storage along with 2.5 million other museum artifacts.

I asked Daeschler the chances of borrowing the professor.

He shook his head in disbelief, smiled, and made a few crosstown phone calls to secure permission from a colleague. John and I then wheeled over to the Museum of Archeology and Anthropology of the University of Pennsylvania to pick up Professor Cope. The curator of collections there had gone to lunch and left the esteemed professor with a security guard at the front desk. Instructions were left for us to sign a

TOP: Professor Cope and his box cushioned with want ads
ABOVE: Edward Drinker Cope
Department of Library Services,
American Museum of Natural History, New York

permission slip like the one you fill out when you check out a library book. That done, the guard unceremoniously slid two small cardboard boxes across his desk and bid us farewell.

The boxed professor didn't weigh very much. He was frail in life, and what was left of him now weighed no more than a few pounds. We whisked the celebrated scientist to our van, eager to make his acquaintance.

We opened the flaps of the first box to find a card affirming that we were being introduced to specimen 4989, better known in life as Professor Edward Drinker Cope. Below the card lay a jumble of bones we took to be the professor's.

The other box had a shipping label from Herbach and Rademan, a company that we later found out is still in business selling electrical parts. We pulled back the unsealed flaps of the cardboard box and found the skull of Professor Cope loosely wrapped in the want ads of an old *Philadelphia Inquirer*. It was disconcerting to see the great professor's illustrious career come to this, but he seemed to be smiling, as skeletons do.

So Professor Edward Drinker Cope, one of the world's most celebrated bone hunters, veteran of the infamous Bone Wars, became our constant traveling companion for the next three years as we assembled objects for our photograph. Through our journeys we were laden with forty-two cases. All of it was pretty much replaceable, except for Ed. What would happen if we lost him? Or if he was stolen or kidnapped? Our next stop was Manhattan, den of thieves. From a phone booth on Interstate 95, I called magazine headquarters to insure the artifacts we had just collected. Over the drone of cars whizzing by, I recited to company lawyers the list of relics and their value. At the end of my list was our newly acquired specimen 4989, which didn't have a price—it hadn't occurred to me to ask the museum what Professor Cope was worth. To insure him legally, our lawyers advised, I simply had to put a value on the item. For a few awkward moments I thought about this. I remembered a junior high school chemistry class where the teacher said the market value of the elements in the human body was slightly over a dollar. But that was the entire body. These were just bones, calcium phosphate, whose price on the open market fluctuates more or less along with the price of fertilizer. What if, for some horrible reason,

the boxes of bones, weighing about seven pounds (3 kg), were lost? I couldn't see myself shipping the museum a ten-pound (4.5 kg) bag of fertilizer with a note saying, "Sorry about Ed, keep the change." Even in his reduced state, he was certainly worth more than plant food. I thought about calling up the museum to see what they valued him at but thought better of it; I didn't want them to reconsider the loan.

We decided to leave a price tag off the professor and not insure him. Even if his past handlers kept him in a cardboard box stuffed with old newspapers, to us he was priceless. But what about his security as we traveled? We decided that while Professor Cope remained in our care it was essential that he didn't leave our side. We found that the boxes fit nicely between the two front seats of the van, and by keeping him beside us we remembered to take them with us wherever we went. While we ate at restaurants he sat quietly in his boxes on a chair next to us. At the hotels at night, while we read books chronicling the professor's legendary exploits, he was tucked cozily between our beds on the night table, still smiling as if he was enjoying the ride. The question we asked each other when we left anywhere was "Do you have Ed?"

BONE HUNTERS MEET THE BONES OF THE GREAT BONE HUNTER

After several weeks, transporting the professor became second nature, like remembering to take the car keys.

At a remote dig site in Utah, Jim Kirkland, paleontologist for Dinamation, mentioned that Professor Cope was one of his heroes. "Really," I said. "Would you like to meet him? He's in the van."

ABOVE: Dinosaur tracker Martin Lockley with the Professor in his backyard near Golden, Colorado.
BELOW: Breakfast with champions: Living legend Bob Bakker dines with Professor Cope at a coffee shop in Boulder, Colorado.

However, as we visited paleontologists and gathered artifacts for the Bone Wars photograph, sooner or later someone would get around to asking, "What's in the box?" Upon our introducing the contents, word quickly spread through the museum that the legendary Professor Cope was in their midst. Whole museum staffs would come giddily over to meet him. Professor Cope was a legend and a personal hero to most of them. We had never realized that bringing Professor Cope to a group of paleontologists was like carting General Patton to a Veterans of Foreign Wars convention. Everybody wanted to meet and be photographed with him—even in his present condition. Quite to our surprise, we took a back seat to specimen 4989, and in a strange way Cope seemed to be escorting us, not the other way around.

SHOW AND TELL AT A BOULDER CAFÉ

One of the most famous bone hunters of modern times is Bob Bakker, known the world over as the gadfly of paleontology. His heretical ideas about dinosaur behavior cast him as the bad boy in the field. Early in our dinosaur coverage I sought Bob's irreverent counsel. We met for breakfast at the North Boulder Coffee Shop in Colorado.

Unbeknownst to Bob, as we eat and talk dinosaurs, the legendary Professor Cope is in his humble box sitting on a chair between us atop my camera bag. I ask Bakker what kind of photographs I should take for the dinosaur story.

"Your pictures have to be counterintuitive," he advises. "Never give people what they expect. Give them what they don't know."

"For instance?" I ask.

"People think that all dinosaurs were big," he says. "There were some that were so small that all they could eat were insects."

"What kind of dinosaur was that?" I say, getting intrigued. "Where can I find it? Who has it?"

Bob tips his white straw cowboy hat back, reaches into his red flannel shirt pocket, and pulls out a small corked test tube of tiny black bones. "I call it *Drinker*," he says, handing me the vial across the table.

Bob is known for bestowing his dinosaurs with unusual names, but not without reason. "Why?" I ask, looking at the little tube of jumbled bones.

"After Edward Drinker Cope," he explains.

I nearly choke on a pancake. When I recover, Bob glances over at the chair between us and asks, "What's in the box?"

MAGIC TOOTH

The Judith River Formation in Montana is great dinosaur hunting ground, but back in the summer of 1876, when Cope and his field crew arrived, it was established Indian territory. The Battle of Little Big Horn, where Chief Sitting Bull and his Sioux warriors had wiped out Custer and his infamous Seventh Cavalry, had just taken place. Upon reaching Montana and hearing the news, Cope's field crew was shaken, but the professor thought their timing was perfect. Sitting Bull and the Sioux, Cope reasoned, would be too preoccupied fleeing the United States Cavalry, which was seeking restitution.

One night Cope and his crew found themselves camped across the river from a teepee village of some two thousand Crow Indians, hereditary enemies of the Sioux. Crow warriors were well known for massacring whites and blaming it on their enemies. At daybreak a small band of Crow chiefs came galloping into the camp. Cope, who wore dentures, was caught by these local dignitaries as he brushed them at his morning toilette. He hastily finished, put them in, and greeted his guests with a great smile. The Indians, taken aback with the ease with which this great white man could remove this important part of his anatomy, asked him to repeat his powerful magic again and again. For the next week Cope's crew was deluged with wild game and fresh fish, offered by Indians who had traveled many miles to see the magic performed by the one called "Magic Tooth."

I had read that the last years of Cope's life seem to have been tortured by headaches. His temper and irritability escalated to the point where he no longer went out in public much but stayed at his home, which doubled as a museum and laboratory. The source of his rage was usually blamed on old age. Now we know better.

Paul Sereno, who found some of the earliest known dinosaur relatives, is not only an accomplished

Paleontologists Dale Russell (left) of Ottawa and Paul Sereno of the University of Chicago are introduced to the legendary Professor Cope. Paul found an anatomical clue to the source of the late professor's notorious headaches and bad temper, an abscessed tooth.

dinosaur hunter but teaches anatomy at the University of Chicago. While visiting fellow paleontologist Dale Russell in Ottawa, we introduce him and Sereno to Professor Cope. Paul inspects the professor's skull closely and exclaims, "He must have been real cranky in his old age!"

I ask why, and Paul points to the gnarled root of a molar cutting deep into the back of the great professor's upper jaw. "That abscessed tooth probably gave him a real bad headache and made him hard to be around," he says.

PROFESSOR COPE VISITS GHOST RANCH

Ghost Ranch, New Mexico, is best known as the residence of artist Georgia O'Keeffe, who found inspiration from its haunting landscape. The ranch got its name from Spanish settlers who claimed to have seen

serpentlike ghosts levitating above its rocky outcrops at night. They were vindicated to a degree in 1947 when paleontologist Ned Colbert of the American Museum of Natural History found a mass grave of some one thousand predatory dinosaurs called *Coelophysis*.

I made the pilgrimage to the quarry, a dinosaur mecca, where I met Lynett Gillette, the acting curator of the Ruth Hall Museum of Paleontology at Ghost Ranch.

As we both have little time, I suggest that we conduct the interview at the famous *Coelophysis* quarry. As I drive my van, she directs me down a secluded gravel road that comes to a dead end at the base of the hills. Even though the Ghost Ranch is now run by the Protestant church, she advises me not to leave valuables in the car. Since my hands are full of camera and recording equipment, I ask her to take the cardboard box. (At this point in my travels, I had thinned the professor down to one box containing his skull.) As she leads me up a hiking path to the quarry, I ask her who named *Coelophysis*. I'm making small talk. I had assumed that Edwin Colbert, who discovered the site, had named these ten-foot-long (3 m) dinosaurs.

"Edward Cope named them," she says. "Have you heard of him?"

I stand in the trail, stunned. As wonderful as it would be to reveal the contents of the box under her arm, I can't bear to do it. After all, we have just met and we're alone in the woods and I'm afraid she'll freak out—hell, I'm afraid *I'll* freak out.

"I've heard so much about him I feel like we're close friends," I answer.

With the box firmly under her arm, she smiles back at me and says, "*Coelophysis* was discovered in 1881 by David Baldwin, who worked for Cope. Cope himself never visited the ranch."

"I see," I say.

When we arrive at the quarry near the top of a hill, I tell her I'm unfamiliar with her work, and she informs me she is more of a geologist than a paleontologist and apologizes that her bone work mostly involves modern syphilitic bones.

The coincidences are too perfect. Removing the professor's skull from his box, I say, "I have to introduce you to somebody. Can you confirm if my friend here had syphilis?"

She tells me the long bones are more diagnostic for detecting the disease and then asks suspiciously, "Who's your friend?"

She doesn't freak, but it takes me an uncomfortably long time to explain how she came to be carrying Ed on his first visit to Ghost Ranch.

Later I learned the surprise visit of the distinguished Edward Cope warranted an article in the Ghost Ranch newsletter. It was, after all, the first time he had been there.

THE COPE AND MARSH REUNION AT YALE

Yale was the only institution that would not part with its artifacts, so we had to photograph our still life of artifacts at the Yale Peabody Museum, the repository built by Cope's archenemy, O. C. Marsh.

As John and I walk into the front lobby of the museum, the boxed Professor Cope tucked under our arms, I wonder if there are any spiritual dues to be paid for the irreverent visit we are initiating. In the foyer as we wait for John Ostrom, head paleontologist

RIGHT: Ghost Ranch curator Lynett Gillette with the boxed Professor at Ghost Ranch *Coelophysis* Quarry.
OPPOSITE: Although Cope had named *Coelophysis*, a dinosaur discovered near Ghost Ranch by his collector David Baldwin in 1881, he had never been there himself until we made our pilgrimage in 1992. Cope is pictured here behind the home of famed artist Georgia O'Keeffe.

ABOVE: Yale Peabody Museum, haunt of O. C. Marsh
OPPOSITE: A postmortem reunion of bitter rivals Cope and Marsh. In their celebrated feud, known as the Great Bone Wars, 136 new species of dinosaurs were described.

at Yale and Marsh's modern counterpart, to escort us through his archives, John and I begin muttering apologies into the boxes.

Professor Ostrom spots us as he comes down the steps. After introductions he asks curiously, "What's in the box?"

I reach into the box I'm carrying for our ever-smiling traveling partner. "Professor Cope, I'd like you to meet Professor John Ostrom of Yale Peabody Museum."

Professor Ostrom jumps back as if I had just unpacked a rattlesnake. "My God!"

"You can call him Ed," John says matter-of-factly. "We do."

Speechless, Professor Ostrom approaches my outstretched hands offering Professor Cope's skull. His emotions seem to dance between horror and fascination. "I wonder if I'll ever be forgiven for this," he says, receiving his distinguished guest in his hands.

"Probably not," I say. "You should show him around, though. Probably didn't get a chance to visit here much when Marsh was around."

While Ed was at Yale, a few strange events occurred that gave us all a little pause. We photographed the still life in the middle of the night in the basement of the museum. About three o'clock in the morning our Polaroids, used for lighting checks, started disappearing, and later we learned all of our 4" × 5" film was mysteriously loaded backward, rendering all our work useless.

On graduation day at Yale, when we photographed Cope with an oil painting of Marsh done from a live sitting, almost all of our photographs have a weird unaccountable blue glow around Professor Cope's skull. Then while we were photographing Professor Ostrom in the main hall of the museum, a light next to Cope's bone box exploded into flames. Cope, however, seemed to be smiling through it all.

COPE RECEIVES HIS DEATH WISH

A year after I introduced Professor Cope to Dr. Bob Bakker, I was back at the same restaurant with Bob when he mentioned some of his new research. *Homo sapiens*, he said, one of the best-known primates, quite surprisingly still lacked a type specimen. It seemed that Carolus Linnaeus, the father of modern taxonomy who named our species in 1758, was content with a short Latin description which means, translated, simply "Know thyself." But the ruling authority on new species, the International Commission on Zoological Nomenclature (ICZN), which was founded in 1895, declared that all species, to be valid, had to have a scientific description and have a registered type specimen at a recognized museum. Bob had done a search of the current literature and hadn't found either a description or a type specimen for modern man. Now he proposed that we nominate Professor Cope for the vacant post. I enthusiastically concurred.

Bob told me we were actually designating the professor to the status of what the ICZN's codebook calls a lectotype, or the "elected type specimen," a term used when the original author of a named species, in this case Carolus Linnaeus, did not select a type specimen.

Technically we could have changed the name *Homo sapiens* to anything we wanted, which set my mind reeling, but Bob is a scientist who wants to simplify science not complicate it. "Just a bit of simple taxonomic bookkeeping," he laughed.

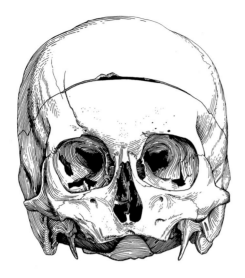

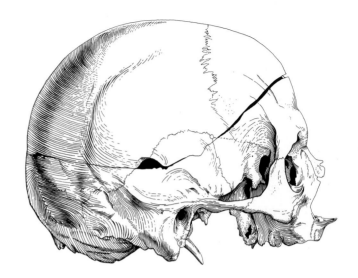

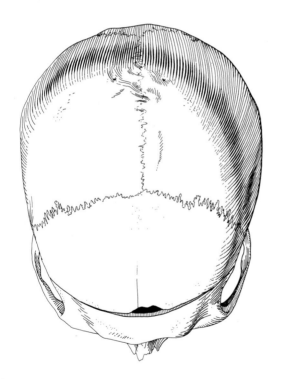

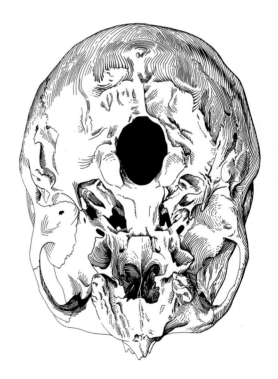

Drawings of Cope's skull by Pat Redman

The scientific description was actually quite straightforward, attaching some numbers to "Know thyself" in terms of brain size. We had to first distinguish our representative Professor Cope from our nearest extinct relative, *Homo erectus*, which has about a one-third smaller brain, by making a volumetric study of the professor's braincase. We accomplished this one memorable Saturday afternoon in November of 1993 by pouring measured doses of acini pepe, a small beadlike pasta, from a graduated cylinder into the professor's skull and rapping the side of his skull slightly to settle the contents. I logged in the data and wrote up the notes of the experiment into Bob's field diary, and he submitted the paper to a dignified but amused review board. In 1994, almost a hundred years after his death, Edward Drinker Cope got his wish and was entered into the scientific literature as the elected type specimen for *Homo sapiens*.

As important to scientists as the type specimen is its "place of capture," which in this case was Philadelphia, Cope's home, where we picked up the box. The city of brotherly love and mustard-coated pretzels is now the registered type locality for humans. Returning Professor Cope to the museum in a cardboard box now seemed highly inappropriate for his exalted new station, so John, who is also a fine woodworker, crafted a mahogany box lined in red velvet and fitted it with a brass plaque.

Professor Edward Cope had been reduced to bone, catalogued, boxed, and shelved like one of his numerous specimens. But in the caring hands of all those admiring paleontologists, one of whom made him the ultimate example of *Homo sapiens*, he seemed to be smiling, as if death wasn't so bad after all.

TOP: Professor Cope, the newly elected example of man, in his new home.
CENTER: Cope's nameplate
BOTTOM: Bob Bakker pouring pasta into Cope's noodle for a volumetric reading of *Homo sapiens'* brain size, to compare with that of our next of kin, *Homo erectus*, a species which had about a third less cranial capacity.

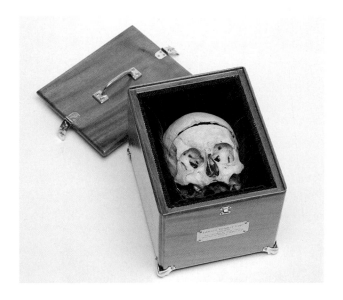

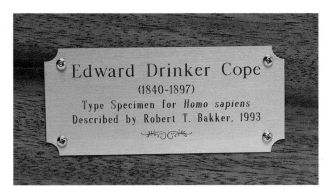

THE TRIASSIC

IN EVOLUTION, SOMETIMES YOU'RE A FLY, SOMETIMES YOU'RE THE WINDSHIELD

There have been several major destructive extinctions along evolutionary history that enabled some groups of animals to leapfrog onto center stage. The extinction usually associated with the demise of the dinosaurs at the end of the Cretaceous Period was relatively modest compared to the one that ushered in their rise at the end of the Permian. It has been estimated that the great Permian extinction, some 245 million years ago, destroyed up to 90 percent of existing life-forms.

Although the archaic forerunners of mammals persisted until the Jurassic, these big, hearty predecessors diminished during the Triassic and were finished off by the end of the Jurassic. Mammals, when dinosaurs took over, fell into the lowly roles historically allocated to rat-sized critters. They were banished to the dark ages for nearly 135 million years, coming out only at night. As early as the end of the Triassic, dinosaurs ruled the day.

Very little of the land fossil record from the beginning of the Triassic is preserved, but what can be scrounged together reveals that some reptiles were making steady innovations in locomotion, becoming more upright and maneuvering on two hind legs, which allowed their forelegs to become arms, perhaps enhancing their ability and efficiency to capture prey. In short, some reptiles were becoming dinosaurs. It was an experiment of nature that proved fruitful. By the end of the Triassic, dinosaurs had evolved and diversified and flourished until they ultimately had taken over all the major roles of land animals.

Metoposaurs, early rulers of the earth, lounging in the Triassic swamps of what is now Petrified Forest National Park.
Illustration by Douglas Henderson, from Dawn of the Dinosaurs

DINOSAURS BECOME WINNERS BY DEFAULT

THE LAST BIG PARTY WEEKEND OF THE TRIASSIC

As Kevin Padian strides across the baked badlands of the Petrified Forest, volcanic ash crunches under his hiking shoes as if he were walking atop the crusted topping of a *crème brûlée*. When he spots a trail of bone spilling out of a hill, his pace quickens and his eyes dart across the ground as if he were tracing the climactic pages of a mystery novel. We are in an area of the park where the rangers discourage hikers. An estimated thirty thousand pounds (13,620 kg) of petrified wood walks out of the park illegally every week, but we're searching for valuable treasures overlooked by the park's more unscrupulous visitors.

Kevin Padian, professor of integrative biology and curator of paleontology at the University of California at Berkeley, is one of the world's authorities on the Triassic/Jurassic boundary, a period of time 208 million years ago when one group of strange and wonderful archosaurs surrendered the earth to another, the dinosaurs.

Kevin falls to his knees. "This is great stuff!" he says as he picks up some blue shards of bone scattered on the surface. He then adds, "We're only about a thousand years too late."

ABOVE: "Looks like this one swallowed a hand grenade," said Kevin Padian, professor of Integrative Biology at the University of California at Berkeley, as he inspected the scattered remains of a metoposaur in the badlands of Petrified Forest National Park.
OPPOSITE: Pistol-packing paleontologist of Petrified Forest National Park Vince Santucci with former Triassic gang members the archosaurs. Armor-plated vegetarians called *Desmatosuchus* (left) and meat-eater *Postosuchus* (right) ruled this once-forested real estate before dinosaurs took over.

I kneel next to him on the top of a knoll. "What is it?" I ask, looking at the thousands of tiny weathered scraps of bone surrounding us.

"I'm not sure yet," he says, overwhelmed with the poor preservation of his weathered creature. "Looks like this one swallowed a hand grenade." He spends the next half hour in deep concentration trying to find a couple of warty, textured pieces of his reptilian jigsaw puzzle to fit together so he can make an ID. Finally he announces triumphantly, "It's armor from a metoposaur, a giant toilet-seat-headed amphibian."

Suddenly John gives an excited shout from the top of a neighboring hill. We abandon our Triassic Humpty Dumpty and race to inspect John's find where some recurved teeth lay about the barren ash like discarded kitchen knives.

"It's a phytosaur," Kevin says, running his finger down the serrations on the leading edge of a large tooth. As his inspection progresses, he amends his diagnosis. "It's a *giant* phytosaur."

I ask Kevin, who lectures 650 neophyte dinosaur students every year, if he thinks the beast, a large meat-eater, would have fought with dinosaurs.

"Predators are animals that don't make their living by going after the big bull males of their species, a

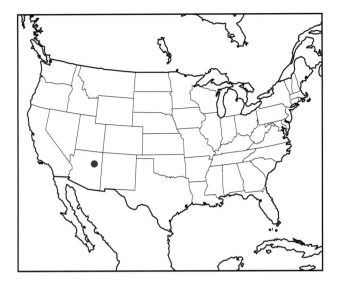

Site map of Petrified Forest National Park, Arizona

competing species, or a prey species," he says. "They go after the old, the halt, the sick, the lame, or the very young. Like lawyers."

Padian's motivation to become a paleontologist was not simple. "For me, the drive is not to describe this neat new fauna and add to the phone book," he explains. "The reason for doing it is to be a thinker. Dinosaurs can provide you with a lot of answers if you ask the right questions. I'm interested in big-scale evolutionary questions. You can look at fruit flies all you want and learn a whole lot about evolution. But there are some things a fruit fly will never be able to tell you—like how major changes in evolution get going. The Triassic/Jurassic boundary is where a major evolutionary change was taking place. In the Triassic, dinosaurs are not generally the dominant part of the fauna. They aren't that abundant or diverse. Dinosaurs were just starting to evolve."

With the concentration of a bomb defuser, he begins delicately removing dirt and rock from around the protruding bones of the ancient beast. He continues, "The end of the Triassic is the busiest time in the history of land vertebrates. It was the last big party weekend, and what was left was the whole basis for the modern fauna on land. Everyone was there except the birds—there were crocodiles, turtles, lizard relatives, pterosaurs, the Permian holdouts like therapsids,

giant amphibians, and even the first mammals, but dinosaurs went on to rule the world. Back then, the smart money would have been on crocodile relatives, not mammals. The mammals were all rodent-sized."

The party weekend was a 5-to-10-million-year affair after which the dinosaurs left about every other archaic archosaur gasping under the table. By the beginning of the Jurassic Period, dinosaurs started conquering almost every available niche. In their ecological coup they filled all the major and supporting roles of the animal kingdom, from large meat-eater to big plant-eater. Their stunning and relatively sudden success is a hotly debated topic in the scientific community.

"There are two basic models that are pushed for their success," Kevin says. "One is competition—the other is opportunism. I don't think you can really tell which it was. I think there may be other ways to look at it. I'd rather focus on what made dinosaurs different from other animals."

Kevin believes it might have been the dinosaurs' specialized anatomy that gave them the edge over their competitors.

"Dinosaurs and pterosaurs couldn't sprawl like crocodiles. They had to walk upright because of the way that their hips are put together. And walking on their hind legs freed their forearms to grasp prey. Nobody else did that before. They also have an ankle that flexes back and forth. It's got two double roller joints on it. Just like a chicken leg. The drumstick."

Viewed against the larger backdrop of evolutionary history, he says, our unassuming dinner fare like chickens and turkeys are missing links to their more exalted past. Birds, he claims, are living dinosaurs.

"Feathers are basically elaborated scales," he says. "In fact, if you look on a bird's leg you will see scales right down by the foot. It's commonplace. You won't find feet on chickens at the Safeway. They cut them off—Americans think they are disgusting. You have to go to the Chinese markets to find chicken feet. In my classes at Berkeley I try to hit the students with the idea that birds are dinosaurs right before Thanksgiving. Then they can go home and point out all the disgusting features on the turkey carcass so they can gross everybody out. It also gets everybody thinking about dinosaur biology.

"Our picture of the Triassic/Jurassic boundary is getting better because we have better methods of attacking problems and gathering evidence. We have better radiometric dates, pollen dates, and dating of microvertebrates. We have better intercontinental correlations. Better data on climate and paleoecology. Probably a lot of it is wrong, but a lot of it is better than it was before. At the end of the millennium, the end of this century, we'll even have a better picture of how dinosaurs evolved. The rate at which new dinosaurs have been discovered has gone up precipitously—tremendously. And with the new discoveries of dinosaurs comes a better understanding of their diversity and how their relationships fit together."

Dinosaurs, Kevin goes on, should give us humans pause. "It's important for people to see that there is always a replacement of fauna, that the dinosaurs weren't just marking time until something brighter and more successful came along—like yuppies. Dinosaurs were animals that were tremendously successful for a hundred fifty million years."

DAWN OF THE RAPTOR
PAUL SERENO'S SEARCH FOR THE FIRST DINOSAUR

Paul Sereno, at age thirty-five, is perhaps the brightest young light in dinosaur paleontology today. He is known within the scientific community as the paleontologist who found some of the earliest known dinosaur relatives.

His dedication to paleontology belies his troubled youth. One of six children in an intellectually driven family from Aurora, Illinois, Paul was slow to find interest in anything but causing trouble. At school he refused to learn to read, he tossed apples in the school band's tubas, and on one occasion threatened to throw a student out of the school's third-story window. Paradoxically, it was his delinquent behavior that turned his life around. One day at the school library, he decided that the school owed him a book. Not just any book but the book that he liked best out of the entire library. Shelf by shelf, he diligently logged and systematically graded the attributes of his future acqui-

sition until he found a book that stopped him. It was Carroll Fenton's *The Fossil Book*, full of strange and mysterious extinct animals that captured his imagination totally and turned him around completely. Well, not completely. He still has the book—it's in his office at the University of Chicago, where he teaches anatomy.

Paul Sereno is articulate and contagiously passionate about his work. He is the recipient of a Packard grant which he is using to break new ground with dinosaur cladistics, "the cookbook method for family trees," a way of charting evolutionary change with so many features that they need computer programs to log details of the animals' anatomy. The computer then systematically reveals which animals may be related by their shared specialized features.

When Paul embarked on this high-tech pioneering mission of charting the dinosaur lineage, he found something very startling—nobody really knew what a dinosaur was. A useful scientific definition just didn't exist.

Everyone agreed that a *Brontosaurus* and a *Stegosaurus* were dinosaurs, but when you started to ask which

Paul Sereno

ABOVE: The elusive *Herrerasaurus*, a meat-eating dinosaur
ancestor from the Valley of the Moon.
Illustration by Shannon Shea
LEFT: Ischigualasto, Argentina—"Valley of the Moon"

few characteristics define a dinosaur, the answers got hazy. A large part of the problem was that nobody had found a complete specimen of an elusive creature called *Herrerasaurus*, reputed to be one of the earliest dinosaurs, sort of the Adam of dinosaur creation. There is only one good slice of real estate left that preserves this *Herrerasaurus* Garden of Eden, which existed some 230 million years ago, and it is in the remote region of northwest Argentina near the Andes. Some of the best collectors in the world had been through this hallowed ground and after years of work had only come up with a few miserable scraps. This Eden didn't give up her dead easily. In 1987 Paul went to some South American museums to investigate these scraps and realized that no matter how long he stared at them, it wouldn't tell him what the earliest dinosaur looked like. Incredibly, the dinosaur family tree was without roots, and his research was foundationless unless he could answer this very basic question: What is a dinosaur? He realized he would have to reopen this mysterious chapter of the Triassic himself.

His grant reviewers, with sage wisdom, warned that pursuing the elusive *Herrerasaurus* would be a fruitless quest, like searching for the Holy Grail. Against all reasonable odds, with a couple of dedicated graduate students, a few thousand dollars, and a ramshackle jeep, he journeyed to the past dinosaur paradise known to tourists as "Valle de la Luna" or

"VALLEY OF THE MOON"
and
"THE BERMUDA TRIANGLE OF EDEN"

After an eight-hour drive from San Juan, Argentina, on some of the worst roads in the solar system, the Valley of the Moon has nothing to offer a tourist except what the name implies. Through some clever marketing, Swiss and Germans are lured there by the busload. They pile out of their air-conditioned coaches, clutching guidebooks and cameras, and for a few brief moments take hundreds of pictures of each other in front of signs and then climb back in, confident that they documented their visit to one of the most desolate landscapes on earth.

Paul's camp was pitched among the bushes at the bend in a dried-up riverbed. Revelry began just after sunrise as Cathy Forster, Paul's postdoctoral student, practiced the banjo in her tent, rousing the paleontologists. Water was scarce, so showers and shaving were abandoned, giving the campers the look of deserters in a badly financed war. Clothes spread out over bushes to dry had collected dew overnight. Drying meat and athletic socks hung from clotheslines strung from trees. A couple of large Sears screen tents served as camp headquarters and as a refuge from the flies and mosquitoes. Paul, the first to rise, was diligently plotting the day's activities on a crate of fossils. Someone coaxed a fire from the embers of the previous night's bonfire to start a pot of coffee, and while the group waited for the first round of caffeine, another informal line formed to await return of the camp's latrine shovel. Breakfast was self-service, and most people stood around a small barrel of caramel-like jam, dipping crackers into the goo and washing it down with cups of thick black coffee.

We had heard the camp was dangerously low on provisions, so we stocked up. In life, there's always time for suffering, but this wasn't one of them—we were on an expense account. On this first trip to the valley we arrived with two all-terrain vehicles full of food, and a few weeks later, while collecting aerial photographs of the area, we helicoptered in emergency supplies to the unsuspecting camp. When we descended, our prop-wash spread yet another coat of dust on the camp below and sent latrine-goers pulling up their pants and ducking for cover. Still, we were welcomed like the Red Cross. We came with slabs of meat, smoked salmon, fresh olives, canned oysters, foie gras, loaves of bread, cases of wines and beer, champagne, crates of fresh fruit and juices, enough varieties of cheese to satisfy a Frenchman, meat, and lots and lots of chocolate.

The only sign of civilization in the valley is a small coffee shop at the entrance that doubles as a park headquarters. There you can buy small jolts of Argentinean coffee, postcards, and pottery that is sure to break if you buy it before you drive on the bumpy loop road around the park. The park ranger who stands behind the bar sipping strong coffee seems to have one function—informing disappointed tourists that there are no craters.

More accurately, the Valley of the Moon looks like the red planet, Mars. A several-thousand-foot escarpment bordering the valley is composed of the same dehydrated iron oxide that gives the red planet its eerie color.

It was the middle of the day, and the light was bad for photographs, so I split off from Paul's excavation to prospect for dinosaurs. I walked down a maze of ravines, head to the ground, until I suddenly

OVERLEAF: Ischigualasto, Argentina, was the dinosaur Garden of Eden in the Triassic: The Sereno expedition drives through the Valley of the Moon. Although the continents were joined in one landmass known as Pangaea during the mid-Triassic, only the Valley of the Moon in Argentina is known to have preserved fossils of dinosaurs from this slice of time, which marks the advent of dinosaurs. In spite of being initially discouraged by grant reviewers, Paul Sereno and his colleagues have had unprecedented success in discovering clues to the origins of the earliest dinosaurs.

DINOSAURS BECOME WINNERS BY DEFAULT

noticed some reddish fragments of bone. My heart raced. I traced the bone up the side of a hill to where part of a skull and some teeth were spilling out of the ground. Slowly I brushed away some dirt from the back of the neck and carefully revealed the sequence of vertebrae going back into the hill. The animal was intact. My heart was now thumping audibly in my chest. I grabbed some stones and scrambled to the top of the hill and made a small pyramid to find my way back to my treasure. As I sprinted back to get Paul, I was already thinking of names for my animal: *Louiesaurus*, *Psihoyos rex* . . .

"Congratulations," Paul says a short while later, inspecting the teeth of my animal. "You found another bloody rhynchosaur."

"Is it rare?" I ask, hopefully.

"In the Triassic they were about as common as sheep are now," he says. "We get a little bored with them. We'll plot his position, but we won't collect him. We have dozens of them."

As I walk back down the hill, my dreams of *Louiesaurus* vaporize. I spot another bone, pick it up, and with reserved enthusiasm ask Sereno if he can identify it.

Paul inspects it with a smile. "It's the distal end of a herrerasaur femur. This is great. We'll be able to plot his size and compare him to the other herrerasaurs."

Paul tells me the story of how he found a complete *Herrerasaurus* on the previous expedition. "We were trying to totally prospect this lower corner of the valley which had a rich pocket of fossils. I climbed up this ridge and saw a little ravine that didn't have any of our footprints, but there wasn't time left that day and we were moving up the valley. I went back and joked to the crew, 'There's a good little triangle back there. It's the Bermuda Triangle and our dinosaur's in that area.' I couldn't sleep thinking that we had left one triangle uncovered. A few weeks later we took a Sunday off to take field pictures and I said to the crew, 'Well, we've got a half day left, you don't have to come, but I really would like to drive back.' Most of the team came. I was up and over the hill and down in the valley first. Cathy Forster was to my right about a hundred feet [30.5 m] behind me. We each went into our own little ravines to prospect. I laid my backpack down on a ridge where I could see it. And I walked about twenty feet [6 m] and saw this fossil coming out of the rock. And I thought, *Well, it's just a rhynchosaur. It's nothing. It's gotta be a rhynchosaur. It's not a herrerasaur.* And I looked at it and then I began to recognize neck bones. And those neck vertebrae were very long. Then I realized I was looking at the whole fossil. I was frozen. I saw the back of the skull and it was a dinosaur. There was no question about it. It was diving into the rock. It was a one-in-a-million find. Here was THE fossil! And I just let out the biggest yell . . .

'YAHHHHHHH! WE FOUND IT!'

"And Cathy ran up and said, 'Either somebody died or you found it.'

"I couldn't stand to look at it. I was afraid it's going to disappear or turn into a rhynchosaur, so I just walked away as people were assembling around this thing. And then I walked back, and what I didn't realize was [that] there was a tremendous release. I was a young professor and this was the first expedition I had led in my life, the first expedition to South America, the first real grant I had received from NSF [National Science Foundation]. There was a lot of pressure. I had grant reviewers saying, 'You can't do it. You won't be able to do it. Sure, go down to South America, but you won't find what you are looking for.' There were a lot of eyes on us. We were reopening the Triassic, and this hadn't been done in thirty years.

"I started crying by the fossil. The crew surrounded me, and the women were holding me, and they were saying they had never seen anybody react to a fossil like that. They said, 'Why are you crying?' and I said, 'Look at this thing. You could have found the toes sticking out and not know you had the skull. You could have found the ribcage, and the skull could have washed away. There's a hundred ways you could find a whole skeleton of *Herrerasaurus* that wouldn't give us what we wanted or we wouldn't have known what we had.' And here it was on a knob of a hill; part of the neck actually was just rolled down a bit and maybe a month later it would be gone. I just rolled the two vertebrae back into place. We could have missed the whole thing. Even if we had found what we came to look for, it could have come a hundred different

Herrerasaur, which terrorized this once-lush valley during
the dawn of dinosaurs in the mid-Triassic some 228 million
years ago, was known only by scattered fragments before
Paul Sereno found one with a beautifully preserved skull,
which he holds above.

ways. To have it come like this, beautifully preserved right on the surface, in the Triassic. . . . It was perfect. It was one of the greatest moments in my life."

In the three years since Paul discovered his *Herrerasaurus* he had accumulated a better understanding of its life. "*Herrerasaurus* was a very active predator, the largest bipedal animal that the earth had ever seen. I don't know that it means all that much to walk on two legs rather than four, except that you have given yourself a new way to capture animals. It balanced on its hind legs, giving it a little agility, so it could manipulate its forelimbs in a humanlike way, for grasping. *Herrerasaurus* was very hard to align with anything that came afterward because it was so primitive, so close to the common ancestor of all dinosaurs. It had fifteen or sixteen features which all other dinosaurs had. We had the whole skeleton, so we could then look at the hand and skull for the things that separated this dinosaur from others. To our great surprise, it had a few features of one of the particular dinosaur lineages, the theropod [meat-eating] dinosaurs. It had enormous hands almost half the length of the forelimb—much like those which characterized theropod dinosaurs for the next one hundred fifty million years. It also had a strange, flexible jaw joint. Once the animal had grasped its prey and managed to get it halfway in its mouth, its jaw would wrap around the prey so it couldn't escape."

Paul's quest to solve some of the mysteries of dinosaur genealogy didn't stop with his discovery of *Herrerasaurus*. The purpose of his current expedition to the Ischigualasto Valley was threefold: first, to determine what the environment was like back in the early Triassic; second, because the absolute time interval of *Herrerasaurus* existence was unknown, he needed to get a grain of volcanic ash for radiometric dating; and finally, he hoped to find another dinosaur.

The first part of the mission looked more like some bizarre culinary ritual than paleontology. When we first arrived at the valley, I saw Paul and the team's geologist, Ray Rogers, literally eating a path through the Valley of the Moon. John and I and our Argentinean driver, who for reasons not clear to us liked to be called "Duck," stood watching them from a distance. Every few steps Paul and Ray would reach

down, pick up a pinch of the bleak landscape, and pop it in their mouths. Then they would swish it around, much like judges at a fine wine competition, spit it out, and with brown rings of dirt around their mouths, thoughtfully discuss their sample's attributes as if they were connoisseurs indulging in a particularly fine vintage. They disappeared over a hill, and later they reappeared as specks in the distance, still eating dirt, huddling and making notes. Duck shook his head and spat on the ground, muttering in Spanish as his index finger revolved against his temple, the international sign language for *mucho loco*.

"By biting a piece of rock," Paul explains later at camp, "and listening to the noise of the sediment grind against your teeth, you can tell exactly how much sand is in there and what the composition is, quicker and more accurately than any other way. The bigger the sand particles the closer you are, more or less, to a river basin. One way to get a picture of what the valley looked like back then is to start right at the base and chew your way across."

That's reasonable, I think, and try it. I put a bit of dirt in my mouth and listen to the discomforting sound of what seems like my molars being ground into gravel. After spitting it out, I resolve that identifying the environment that way is an acquired taste.

Paul and Ray dined on a smorgasbord of landscapes. They ate their way through ancient rivers and small lakes, patches of petrified evergreen forests and bogs of *Dicroidium*, which sounds like a dessert sure to deliver intestinal trouble. *Dicroidium*, Paul explained, is low-growing plant life, about a meter tall, that would have provided cover for fierce nondinosaur terrors around then. Paul's census of the valley had turned up members of four main groups: huge shovel-headed carnivorous amphibians, the "mammal-like reptiles" which looked like crocodiles with an upright stance, reptiles which included commoners like rynchosaurs, and finally, the new kids on the block, the dinosaurs.

To get an ash date, you need one perfect unaltered crystal of ash the size of a pinhead. If finding a complete herrerasaur was the Holy Grail, then finding a grain of ash that survived unaltered for a quarter of a billion years would be like trying to find the proverbial needle in a haystack. In his initial request for

The bone of a rhynchosaur weathers out of
the ground.

funding, Paul said he would try to get an absolute numerical date for the valley, but he had no idea if there were ash beds down there, and even if they existed, they might be altered beyond usefulness. Ash beds near dinosaur-bearing deposits are relatively rare.

On their first trek across the valley, Paul and Ray had chewed through miles of sediment and didn't find any evidence of ash beds. Their hopes were temporarily dashed. But while they were measuring how far the 1988 herrerasaur was from the bottom of the formation, Ray stumbled onto pay dirt. By a stroke of luck (and a hammer) he found an ash bed just a few meters below where the herrerasaur was found. They dug about a foot to the purest ash, where a crystal of sanidine, long ejected from a volcano, had settled. A quar-

ter of a billion years later Ray took it as a sample to be passed through some exhaustive radiometric procedures. For the first time humans would know when dinosaurs first appeared. Paul's luck seemed to be holding: his first two objectives were fulfilled, and soon came the third.

Ricardo Martínez, one of the Argentinean members of the crew, was prospecting near a dig when he inexplicably picked up a small irregularly shaped rock. He was about to toss it when a glint on the back of the rock caught his eye. It was the sun shining off some teeth in the stone. On closer inspection he saw that the rock was actually the skull of a small animal. A little bit more digging revealed that the whole body was intact but encased in a hard concretion.

Paul and I were driving back from another dig when Ricardo and some of the other crew excitedly ran up to our jeep with the good news. Paul was down

ABOVE: Site of *Eoraptor* excavation
BELOW: Fernando Novas (left) and Paul Sereno inspect the matrix-encrusted bones of a Triassic "roadkill," which will later be christened "dawn stealer."

at the edge of the small trench in seconds, his nose just inches from the fossil, trying to find the ankle joint, one of the diagnostic clues to the creature's hereditary allegiance. There was speculation then that it could have been a small croc, but even so there was an exuberant exchange of hugs among the exalted crew.

This Triassic treasure was passed around the group. I looked at it closely, but in this unprepared state, surrounded by a hard stone matrix, the animal looked remarkably unimpressive. Paleontologists call specimens at this point in the discovery process "roadkills." The animal was wearing nearly a quarter of a billion years of deposits and needed some cleaning up.

Paul took the dinosaur back to the Field Museum in Chicago for a cleaning by a maestro with a dental pick—a fossil preparator—who would skillfully liberate the tiny dinosaur after removing sediment, grain by grain, under a microscope. Paul's third objective was accomplished. This goose-sized, featherweight dinosaur wound up dethroning *Herrerasaurus* from the title of the most primitive dinosaur relative.

I thought it might be interesting to get the little critter's head examined. We made an appointment to visit Andrew Leitch, a dinosaur researcher in Toronto who was having great success using CAT scanners to nondestructively probe the mysterious inner workings of dinosaur skulls.

In the first-class section of a United Airlines flight to Toronto, a curious stewardess notices the tiny dinosaur skull resting on a wad of cotton on Paul's dinner tray and asks if he has brought his own breakfast. Paul tells her that it's the most primitive dinosaur ever found, news that the scientific community won't have until an official press conference can be called a few months later. The little dinosaur in seat 4A starts to cause quite a sensation among the crew.

"Is it a male or female?" another excited flight attendant asks.

"Can't tell," says Paul.

"Is it a baby dinosaur?" asks another, noting its small size.

"It's the only one, so we have nothing to compare it to," Paul replies.

"How old is it?" asks the pilot. Paul tells him that the dinosaur is about a quarter of a billion years

ABOVE: After several months of grain-by-grain removal under a microscope by a preparator with a dental pick, the skull of *Eoraptor* is liberated from its prison of stone. *Eoraptor*, which means "dawn stealer," was pulled from 228-million-year-old sediment and is the most primitive dinosaur discovered to date.
RIGHT: *Eoraptor* flies the friendly skies first-class to get its head examined by a CAT scanner in Toronto.

ABOVE: *Eoraptor,* from the Upper Triassic of Argentina, is a meter-long (39.37 in) bipedal dinosaur, very close to the common dinosaur ancestor.
Illustration by Michael W. Skrepnick
OPPOSITE: X rays and CAT scanning proved useless in unlocking the secrets of *Eoraptor*'s rock-filled braincase.
Image manipulation by Nik Kleinberg

old. And then, revealing the limited concept of time most humans have, the pilot asks, "Were people around then?"

Paul smiles faintly as he gives the pilot a quick lesson in the history of time. The pilot is visibly stunned. He goes back to the controls of his DC-10 shaking his head.

Afterward Paul tells me, "Dinosaurs, more than any other fossil group, maybe even fossil humans, give people an idea that something, something truly different, existed in the past. It makes them just a slight bit smaller. Most people have no concept of geologic time. They live in yesterday and today. To get your average person to understand that humans didn't exist when dinosaurs existed is a giant leap. In the broadest sense, that's the value of the dinosaurs—they teach of times past and open people's eyes to the fact that we are a fleeting moment toward the end of the evolutionary history of life."

The CAT-scan results from Toronto are less than promising. Andrew Leitch reports that the new dinosaur has a head like a rock. The skull, completely fossilized, no longer shows any discernible difference between the bone and the rock that fills its head. The only detail the scan revealed is the three cavities of the inner ear, which I find impressive; but Paul says disappointedly, "Every animal in the world has three inner ear cavities, except for a hagfish or a lamprey. At least we know it's not one of them."

Back at the University of Chicago, Paul phones the university hospital radiology lab to see if they can CAT-scan the little dinosaur on short notice. We hope

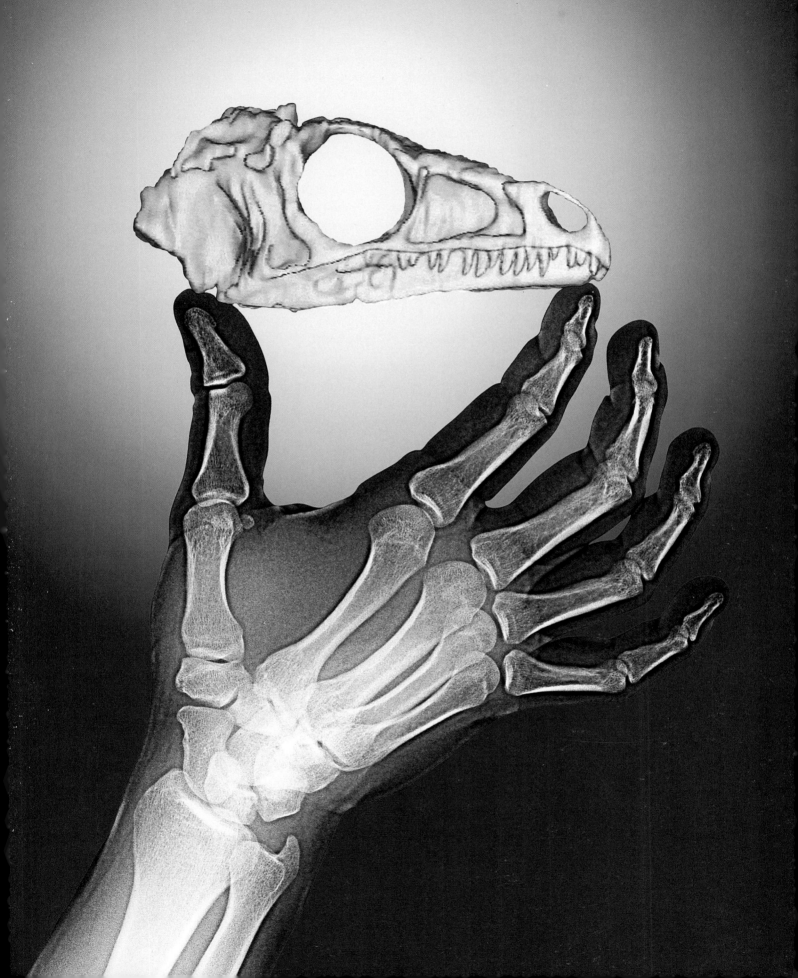

their newer equipment might produce images with finer resolution.

Sitting in the hospital waiting room, Paul describes how the new dinosaur differs from *Herrerasaurus*.

"What the new dinosaur represents is a more ancient lineage. Whereas *Herrerasaurus* was already split off onto the theropod lineage, the new dinosaur had not, even though it also was a carnivore. It had not evolved a jaw mechanism characteristic of theropods. But its teeth are very recurved and very pointed, with small serrations which indicate that it's definitely a meat-eater."

I ask Paul the same question that motivated his quest: "What characteristics define a dinosaur?"

"First of all, size is not among them," he says, "because the new dinosaur is quite small. Principally dinosaurs were getting modifications in the ankle joint and the hip joint, locking the bones together, opening up the hip joint, perhaps to allow them to run more efficiently on their hind limbs."

"Where would you draw the line for the earliest dinosaur?"

"By definition, the first dinosaur would be the common ancestor, the species which gave rise to the two great clades of dinosaurs, the ornithischians and saurischians." All dinosaurs can be grouped by their hip bones into either ornithischian, bird-hipped dinosaurs, or saurischian, lizard-hipped dinosaurs.

"The new dinosaur is very close," Sereno says. "It's just a couple of steps away. It's very rare that we find the actual common ancestor of anything in the fossil record. We're looking at eons of time and on one place on the globe, and when you think about it, the probability of finding an ephemeral species that's only going to last a million years, and maybe lived in a restricted range, is very remote. That we came as close as we did is actually very surprising. Amazing."

One of the problems with sticking dinosaurs in family trees is that you not only have to know every facet of your animal's anatomy but all the others around it. You have to familiarize yourself with an overwhelming amount of anatomy. A contemporary of Paul's calls this "the curse of cladistics."

"His curse is really the beauty and the pleasure of it for me," Paul says. "It's hard work, but this is the whole crux of the matter. It is not so obvious what a dinosaur is. The answers lie in the ground. They lie in museums. They lie in the comparisons that nobody has made. They lie in the mind or the eye of somebody that has the time or the talent to go and make these comparisons from one shelf in one museum to another one thousand miles away from some other part of Pangaea.

"This is the beauty for me as an adventurer, so to speak—to go to collections and find some key piece of evidence that somebody left on a shelf because they didn't realize what it was. I love it. It's wonderful! It's equivalent, in every way, to discovering a new fossil out in the field."

Finally, an overworked nurse calls, "Dr. Sereno." We walk over to the computer terminal where she

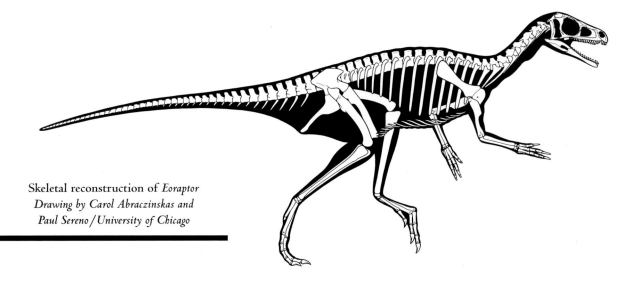

Skeletal reconstruction of *Eoraptor*
Drawing by Carol Abraczinskas and
Paul Sereno / University of Chicago

begins the perfunctory task of initiating the tiny dinosaur's outpatient status.

"Patient's name?" she asks.

Paul at this time is in the middle of deciding the name for his little creature, so he hesitates, much like a parent who is undecided, even after birth, to officially name his child.

"*Eoraptor*," he says to the waiting nurse.

She looks up slowly from her computer terminal.

"It means 'dawn stealer,' " Paul explains.

She gives him a polite smile as if to say, "That's nice."

"Age?" She continues typing, eyes studying Dr. Sereno as if she's checking in an unstable mental patient.

Paul says, "About two hundred thirty million years."

She raises an eyebrow.

"Maybe two hundred twenty-five million," Paul says. "We don't know exactly yet."

The nurse nods in mock understanding. Her fingers poised on the keyboard, she inquires, "Male or female?"

"Don't know that exactly, either."

The nurse scans down the list of examination questions. "Where's the patient from?"

"Ischigualasto," he says, "Valley of the Moon."

Her eyebrow rises again.

"Just put Argentina," Paul says.

The head radiologist, who is acquainted with unconventional patients like fossils and museum mummies, then introduces herself and escorts Paul to the CAT-scanning room, where he places *Eoraptor*'s skull on the table. The table begins to slide forward under the machine's storm of radiation, surrendering the fossil to the forces of modern medical science and technology. While waiting for the results, I ask Paul if there's any similarity between the anatomy of a human skeleton and a dinosaur.

Paul explains, "If I were to stand on all fours, and you were to CAT-scan me and put my image right up next to a brontosaur, you'd have a bone for bone correspondence with the brontosaur. Every bone in your body matches a bone in the brontosaur body. Occasionally there's been a fusion of two bones, like in our human hip region, but every bone matches up. A brontosaur has a few more tail vertebrae, it has a radius and ulna, it has all the digits of the hand, it's got a humerus. It's the vertebrate body plan—it's very conservative. We all descended from animals that have a radius, ulna, humerus. In fact, you can go back three hundred fifty million years ago to the Devonian and look at the fin of a fish called a crossopterygian; you can see humerus, radius, ulna. We're all its descendants."

I ask him if he thinks a paleontologist can teach anatomy better than a medical doctor.

"I think we can. . . . We are able to teach it with more personal interest because this is in fact what we spend a lot of our time studying. Not human bones, perhaps, but bones in general and the way animals evolve. We teach it from an evolutionary perspective to try to get dinosaurs to do for the students what they do for little kids, which is to make them realize that humans are not perfect made-in-God's-image machines. We have remnants of our history built into our bodies that make absolutely no sense. My job is to get the medical student to step back from the animal and look at it for what it is. It's an absolutely bizarre arrangement of things that works most of the time and fails at other times because of the very complicated evolutionary history that put us together. I mean it gave us our jaws that used to be gill arches, and they have their connections because of the way that we used to breathe water through our gills. We have gut walls that fail and we have hernias and back problems because we were at one point walking on all four legs and we're not designed ultimately to walk on two legs. And to really get down to the ultimate questions of 'Why?' with the human body, you have to understand where we came from.

"I don't think that we were created by a Creator, but I do think that paleontologists come as close as we might ever come to playing God themselves because we actually create species that didn't exist before. We just created, out of mere dust, *Eoraptor*. An animal we brought to life. It didn't exist before. We didn't know what it looked like. It had no name. And now we have a beautiful animal. It's got a name and a whole story behind it. We know what it might have done and when it lived. We created *Eoraptor*."

Disappointing news from the university's head radiologist: *Eoraptor*'s skull is still impenetrable to mortals. But a few months later the radiometric lab calls with good news—the dating for the ash from the Valley of the Moon was successful. The Dawn of Dinosaurs can be recorded at 227.87 million years. Plus or minus 300,000 years.

UNSOLVED MYSTERIES AT GHOST RANCH

A red hill doesn't touch everyone's heart as it touches mine and I suppose there is no reason why it should. The red hill is a piece of the badlands where even the grass is gone. Badlands roll away outside my door—hill after hill—red hills of apparently the same sort of earth that you mix with oil to make paint. All the earth colors of the painter's palette are out there in the many miles of badlands. . . . I brought home the bleached bones as my symbols of the desert. To me they are as beautiful as anything I know. To me they are strangely more living than the animals walking around—hair, eyes and all with their tails switching. The bones seem to cut sharply to the center of something that is keenly alive on the desert even tho' it is vast and empty and untouchable—and knows no kindness with all its beauty.*

—Georgia O'Keeffe, January 1939

The artist Georgia O'Keeffe did some of her best work when she lived at Ghost Ranch, New Mexico. And so did many paleontologists. The hills of this 21,000-acre (8,505 ha) ranch contain sediments spanning the three epochs of the Mesozoic: the Cretaceous, the Jurassic, and the famous Triassic red beds of the Chinle Formation—190-to-210-million-year-old sediments that have been a source for paleontological discoveries for more than a century.

"Ghost Ranch got its name from the Spanish people who settled in the area long before the Anglo arrived," Lynett Gillette, the acting curator of the museum there, tells me. "They called it Los Brujos,

* Nicholas Calloway and Doris Bry, eds., *Georgia O'Keeffe in the West* (New York: Alfred A. Knopf, 1989).

which means Witch or Ghost Ranch. They believed that it was a haunted place. There were reports of a serpent that haunted the top of a mesa. Local people would prefer to walk for miles around to avoid coming near this place. The last Spaniards that attempted to live at the ranch sold the ranch to a wealthy Easterner by the name of Arthur Pack who turned it into a dude ranch in the 1930s. Strangely, the first paying guests were vacationing scientists from the Manhattan Project at nearby Los Alamos.

It was during the Pack years that O'Keeffe, like many other well-to-do Easterners, came out to Ghost Ranch and its surrounding badlands for inspiration. But Ghost Ranch's association with fossil collectors came many years before.

David Baldwin, a private collector who prospected for fossils around the region in the late 1800s, was originally employed by the Yale College Museum curator O. C. Marsh, but Marsh wouldn't pay him, so he went into the service of Edward Drinker Cope, Marsh's archenemy. While working for Cope, Baldwin found one of Cope's better-known dinosaurs.

Lynett Gillette tells me, "It was in the winter, Baldwin's favorite collecting time because he could always get water from the snow for his mules. He wandered up a dried-up wash near Ghost Ranch and found some very strange small bones—they were fragile and hollow. He sent them back to Cope, who recognized them as belonging to a dinosaur. In 1881, from these few scraps, Cope named the animal *Coelophysis*, which means "hollow bone."

Baldwin had identified the location of the site as Arroyo Seco, which in Spanish means "dry river."

"Problem is," Gillette tells me, "there are a lot of dried-up rivers around here called Arroyo Seco."

ABOVE: A herd of *Coelophysis,* one of the earliest known
dinosaurs from America, crosses a Triassic stream.
Illustration by Douglas Henderson,
from Dinosaurs, A Global View
LEFT: Ghost Ranch, a 21,000-acre (8,505 ha) dude
ranch and former haunt of famed artist Georgia O'Keeffe,
where thousands of Triassic predators called *Coelophysis*
were discovered by Edwin Colbert in 1947.
OPPOSITE: Ghost Ranch insignia designed by
Georgia O'Keeffe.

DINOSAURS BECOME WINNERS BY DEFAULT

This ten-ton block from the *Coelophysis* quarry at Ghost Ranch was worked on for five years with a microscope and a dental pick by preparator Alex Downs at the Smithsonian Institution in Washington, D.C.

For practical purposes, the site was lost. In the 1920s, paleontologists from the University of Berkeley, led by Charles Camp, found dozens of large reptiles, called phytosaurs, on Ghost Ranch. Although they weren't dinosaurs, these long-tailed, crocodilelike beasts with large daggerlike teeth were found in great numbers and excited the interest of paleontologists. The news also excited the local Spanish people—they had always believed the place was haunted by serpents and now they felt vindicated.

Also interested in Charles Camp's Triassic monsters was the curator of vertebrate paleontology at the American Museum of Natural History in New York, Edwin Colbert. He decided to make a pilgrimage to Ghost Ranch while on his way to explore the Triassic outcrops of Petrified Forest National Park.

I visited Colbert, now eighty-eight years old, in Flagstaff, where he is honorary curator at the Museum of Northern Arizona.

"In the summer of 1947 I had a permit from the government to work at the Petrified Forest," Dr. Colbert tells me. "But first I stopped to do some poking around Ghost Ranch."

With Colbert was George Whitaker, a museum technician, and Colbert's neighbor Tom Ierardi, who had come along for the adventure.

"George Whitaker was on one side of a little canyon," Dr. Colbert recounts, "and I was on the other side with my neighbor when George came across these little bones. He didn't know just what he'd found, so he picked them up and brought them over to me to identify. The original type material of *Coelophysis* described by Cope is at the American Museum. I identified Whitaker's find right away as being from *Coelophysis*. Cope's material was pretty much broken up—really scrappy—but we found many completely articulated specimens."

Dr. Colbert and his crew decided to cancel their plans to explore the Petrified Forest and spent the whole summer removing blocks of stone and bone from Ghost Ranch.

"We had only intended to stay at Ghost Ranch for two or three days, just to explore it, so we set up tents, but when it became evident that we had stumbled

Edwin Colbert, former chairman of the Department of
Paleontology at the American Museum of Natural History
in New York, rediscovered *Coelophysis* at Ghost Ranch with
colleagues in 1947.
"Accidental ingestion" is how paleontologist Dave Gillette
explains the remains of a baby dinosaur found in the
stomach of this *Coelophysis.* "Modern crocodiles will snap
at anything that moves, even their own babies."

onto something big, Mr. Pack, who was the owner of Ghost Ranch, said we couldn't spend the summer in tents. He gave us the use of the Johnson House, a posh Spanish-style summer place built by Mr. Johnson, of Johnson & Johnson, the Band-Aid people. It was the most luxurious field season I have ever had. After a day of digging bones we could jump in a pool, cook meals in a modern kitchen, and read books from the stacks in the large living room."

It was during this field season made in heaven that Colbert met Ghost Ranch's most celebrated resident.

"Georgia O'Keeffe had a house about three-quarters of a mile [about 1 k] down from the main ranch," Dr. Colbert tells me. "She had another house down in Abiquiu ten miles [16 k] down the road, but she spent her summers at Ghost Ranch. She looked *me* up, I guess. She had heard about the discovery, of course, and was interested in what was going on. She came around and I took her up to the quarry. She was interested in bones—she used bones in her paintings. She had me over for supper several times, and we went on a picnic together the following year. People said she was aloof and found her distant, but I got along with her fine. I wasn't trying to be arty—I didn't talk about art—we'd talk about bones. She would ask me to identify bones that she would pick up on her walks; mostly they were from cattle and sheep."

Over the course of the summer Colbert and his small crew had to remove several tons of dirt and rock by hand to get at the dinosaurs. After several large blocks were taken out and transported back to New York, it took several more years to prepare the specimens out of the rock. He got around to finishing the *Coelophysis* monograph in 1989.

"*Coelophysis* was a small dinosaur," he tells me. "On average about six to eight feet [183 to 244 cm] in length. It was one of the bipedal meat-eating dinosaurs. You may be familiar with *Tyrannosaurus* and *Allosaurus*, which are later forms. Well, this was an earlier type, smaller, about seventy pounds [32 kg], but had the same general body plan. We had different ages, from very young ones up to adults—it was a population."

I ask Colbert how it came to pass that thousands of predators were together in one place.

"Well, that's a good question," he says, still mystified almost fifty years after finding them. "Quite obviously, by their sharp, serrated teeth, these were meat-eating dinosaurs, and meat-eaters in our modern world don't travel in big groups. The biggest group of meat-eaters today are a pack of wolves which are perhaps a half a dozen individuals or a pride of lions which may have eight or ten. But here we found thousands!"

When pressed to explain why they were all together, Dr. Colbert tells me, "I don't really know. Maybe they were feeding on fish or something like that. William Bartram, one of the early American naturalists who explored Florida back in the eighteenth century, told of alligators being so thick at one place where he camped that they were almost a solid mass. He said he could have walked across from one side of the river to the other on the backs of these alligators if he had dared. These alligators had all congregated there to feed on a great run of freshwater fish. This may have been what attracted the dinosaurs— I don't know."

Dr. Colbert says that about 95 percent of the specimens at the quarry were *Coelophysis*, the remaining 5 percent were fish. One *Coelophysis* specimen, however, made him realize that this vicious predator probably ate whatever it could sink its teeth into—in the belly of one adult-sized specimen was the remains of its last meal, a baby *Coelophysis*.

"I think *Coelophysis* was a cannibal," Colbert says. "Modern reptiles will eat their young if they get a chance, crocodiles and alligators especially."

The biggest unsolved mystery of Ghost Ranch remains: What killed all the *Coelophysis*?

"Maybe it was a local catastrophe of some sort," Dr. Colbert speculates. "The herd is found in river deposits, so maybe they died while trying to cross a river in flood. You know, a number of years ago a great herd of caribou in Canada tried to cross a river that was running in full force, and about ten thousand of them drowned. Their bodies lined the banks of the river on either side for two or three miles [3 to 5 k] downstream. This herd of *Coelophysis* didn't die very far away, because their skeletons weren't disarticulated. If they died first and then a flood came, their bones would

have been scattered. I doubt we'll ever know why they died for sure. It's like this whole business of extinction of the dinosaurs—it's a good field for speculation, but I don't know how we can ever know for sure."

In 1955 Arthur Pack donated Ghost Ranch to the Presbyterian Church, which uses it as an educational preserve. The *Coelophysis* quarry was reopened in 1981 and 1982 by the Carnegie Museum of Pittsburgh, the New Mexico Museum of Natural History, the Museum of Northern Arizona, and Yale University. The largest block from the quarry is now on view at the newly built Ghost Ranch Museum. This block of bones was so big that the museum had to be built around it. The quarry has been closed until preparators can free the specimens, which Lynett Gillette estimates will take seven more years. Ghost Ranch and its museum are open to the public.

A solitary *Coelophysis bauri* reclines lazily beneath the trunk of a massive Triassic conifer in a misty fern-laden meadow. *Painting by Michael W. Skrepnick*

THE JURASSIC

MOTHER NATURE'S 70-MILLION-YEAR MIDLIFE CRISIS

The Jurassic Period was the middle of the Meso-zoic Era, which means "midlife," and Mother Nature wasn't handling menopause very well. A couple of billion years of manufacturing nothing but amoebas in slime sauce had seemingly made her edgy and overly eager to tap into the limitless possibilities that DNA recipes had to offer. In the Jurassic, it was as if Donald Trump and Dr. Seuss got together and decided to make animals. Everything got big and weird.

The Jurassic ocean over present-day Wyoming was filled with giant sea reptiles, fifty-feet-long (15 m) plesiosaurs that give credence to the Loch Ness Monster. Giant armored squid, Panzer Calamari,* patrolled the oceans with giant sharks. Mother Nature was breeding giant land predators like *Megalosaurus*, which weighed in at ten thousand pounds (4,540 kg). During the Jurassic, Mother Nature conceived the fifty-ton *Ultrasaurus*, those giraffes on steroids, like Kipling's "elephant's child," caught at both ends and stretched out to more than one hundred feet (30 m), larger in size and weight than a blue whale. Stegosaurs, with their mysterious plates down their spines, began to wander along lakes and rivers. Some say this was when some small dinosaurs sprouted feathers, evolved wings, and became birds.

The sediments of this time period reveal the remnants of a very real evolutionary nightmare. Gargantuan monsters controlled the sea, land, and air. These were strange times—Mother Nature had finally become inspired.

*A registered Bakkerian sound bite.

Jurassic Storm by Douglas Henderson

THE REAL BIG-GAME HUNTERS: IN SEARCH OF JURASSIC WILDLIFE

JACK MCINTOSH: GIVING *BRONTOSAURUS* BACK ITS HEAD

When Jack McIntosh was five years old his father took him to the Carnegie Museum in Pittsburgh. He remembers it as the time he first fell in love with dinosaurs in general and sauropods in particular. Sauropod is the correct name for the long-necked, long-tailed animals that are popularly called brontosaurs. But back then, in 1928, *Brontosaurus*, arguably the most famous dinosaur in the world, didn't have a head. In fact, the true identity of this faceless giant was unknown for more than a century after its discovery. Then Jack became the world's expert on sauropods and returned to the Carnegie Museum and gave this dinosaur its rightful head.

There is probably no other land animal more inconvenient to study than the sauropod. Its descendants are spread on just about every continent. Sauropods are a terrible nuisance to move around. They are too large to be sent to laboratories for study, so one must take oneself to the dinosaurs. Although their bones can weigh tons, some of them are incredibly fragile.

ABOVE: Mark Norell, assistant curator (left) of the American Museum of Natural History, removes a *Camarasaurus* head from an *Apatosaurus (Brontosaurus)* mount in 1991, correcting a century-old error.
OPPOSITE: Jack McIntosh, the world's leading expert on sauropods, admires the famous Carnegie Museum of Natural History's *Apatosaurus louisae* in Pittsburgh. This forty-ton vegetarian, named after Andrew Carnegie's wife, is over seventy-seven feet (23 m) long and is the longest mounted dinosaur in the world.
OVERLEAF: When Jack was a child, this *Apatosaurus* was headless, and then for nearly fifty years it was on display with the wrong head, that of a *Camarasaurus*. After becoming the world's expert on sauropods, Jack returned to give the most famous giant in the world its proper head.

They are packed away in the dreariest rooms of the museums: corners of dark, wet basements and noisy, clanking boiler rooms, in dusty, musty dungeons and under stairways—anywhere their discoverers can hide them. Some say that these are the only convenient places to store the giant bones; others say that here they can be conveniently forgotten, because preparing and mounting a sauropod can consume an entire career and cause a lot of back problems. To the handful of sauropod experts, it is not surprising that there are many more unmounted sauropod skeletons in the world than mounted ones. For this reason, Jack McIntosh has never actually gone into the field and dug one up himself: he's too busy keeping up with all the discoveries hidden away in the dark corners of museums, rediscovering, if you will, what's been dug up before.

When Jack was at Yale in 1941, he used to work in the Peabody Museum laboratory putting sauropods back together for fun. He had a great time doing it, he says. Torn between vertebrate paleontology and physics, he chose physics as his career and paleontology as his passion. Not surprisingly, his memory

ABOVE: Dr. Hermann Jaeger, director of the Humboldt Museum, in the basement with bones excavated from the Tendagura Quarry in Tanzania.
BELOW: Carnegie Museum of Natural History in Pittsburgh
OPPOSITE: At the Museum of La Plata University in Argentina, paleontologist Fernando E. Novas stands next to a femur of *Antarctosaurus*, a giant titanosaur sauropod of the Late Cretaceous (70–80 million years old) which may have weighed up to fifty tons.

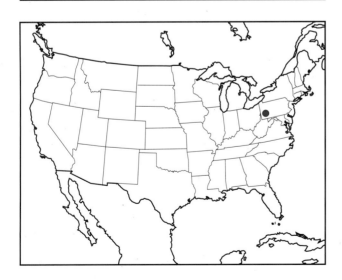

is formidable and his command of research material astounding. He retains a bewildering catalogue of facts, places, dates, and names of discoveries and seems to remember where every sauropod bone in the world is hidden. "The femur is under a table in the basement in La Plata, and unless you just happen to spot it, you wouldn't see it," he says about the biggest femur in the world, a bone I managed to miss twice while visiting this Argentinean museum.

Bob Bakker, one of the most heralded paleontologists, calls McIntosh one of the great unsung heroes of paleontology. But when I approached Jack for an explanation, he struck me as a person who didn't wish to be praised. He is quiet, reserved, shies away from publicity, and dislikes speculation, viewing dried ink perhaps as fossilized thoughts to be scrutinized by future generations. When his opinions varied sharply from those of his colleagues, he politely asked not to be quoted, not wanting to embarrass others. But this publicity-shy professor seems destined for the paleontological limelight over several controversies, the most notorious when he gave one of nature's most celebrated creatures its crowning glory.

MAN OF STEEL CLONES A DINOSAUR

Andrew Carnegie is known as the business mogul who applied Darwinian logic to his business practices and made carnage of competitors in his rise to the top of the steel industry. But at least among paleontologists he redeemed himself and is regarded as the patron saint of dinosaurs. When Carnegie was erecting the Carnegie Museum in Pittsburgh, he wanted his Dinosaur Hall filled with the biggest and best dinosaurs money could buy. In 1909 a great bone bed was discovered near the little town of Jensen, Utah, and it was to become the biggest source of Jurassic dinosaurs anywhere in the world.* Carnegie financed this quarry and ran it like one of his factories, employing dozens of men working year-round, even through the harsh winters. Over the years the Carnegie Quarry, as it became known, produced more than four hundred

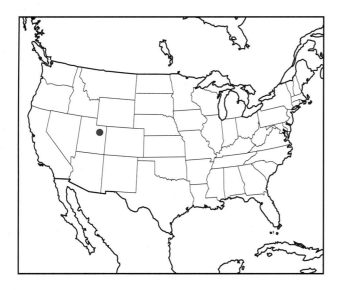

ABOVE: The Carnegie Quarry, now called Dinosaur National Monument.
RIGHT: Paleontologists have chiseled the remains of several hundred Jurassic dinosaurs from their rocky tomb since work began in 1909 at what became the Carnegie Quarry near Jensen, Utah. The site, discovered by Earl Douglass, later became known as Dinosaur National Monument. A permanent building was put over the site in 1958 to preserve *in situ* the bones, which attract nearly 500,000 visitors per year.

---

*Daniel J. Chure and John S. McIntosh, "Stranger in a Strange Land: A Brief History of the Paleontological Operations at Dinosaur National Monument," *Earth Sciences History* 9, no. 1 (1990).

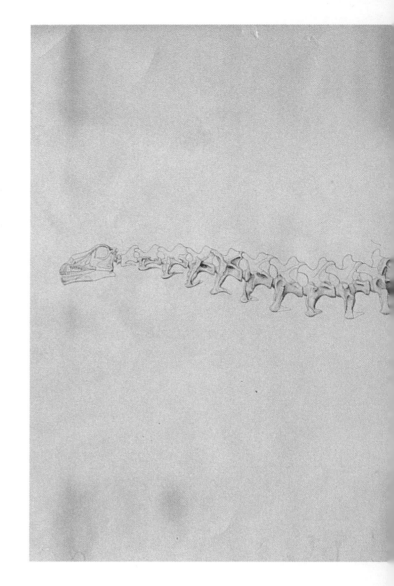

ABOVE LEFT: Andrew Carnegie and his wife, Louise, circa
1917; both had dinosaurs named after them.
Carnegie Library
ABOVE : Marsh's reconstruction of a *Brontosaurus* with the
dotted outline of a *Camarasaurus.*
Smithsonian Institution

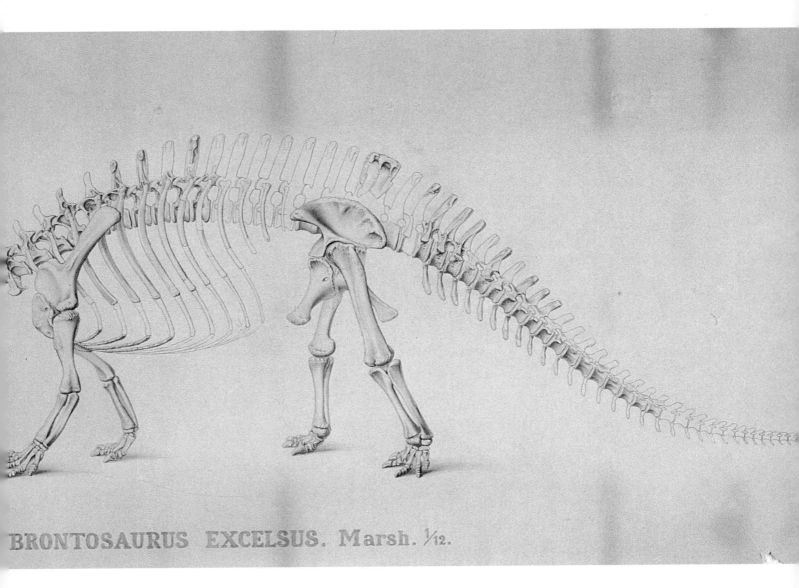

BRONTOSAURUS EXCELSUS. Marsh. 1/12.

dinosaur specimens representing some ten genera of dinosaurs and twenty mountable skeletons.*

Carnegie was so proud of a large sauropod named *Diplodocus carnegii*, kind of a thin, dieting brontosaur with a long, whiplike tail from the Sheep Creek Quarry in Wyoming, that he had it cast and bequeathed to ten museums around the world. But the most celebrated creature to be resurrected from Carnegie Quarry was undoubtedly his *Brontosaurus louisae*, named after his wife. This brontosaur looks like a larger, much fatter version of *Diplodocus*, but like most sauropods, it too was discovered headless, or so it seemed.

Like a little joke of Mother Nature's, a paleontologist will nearly always find a sauropod butt first and
*Ibid.

only after several years of hard-rock mining finally discover that the beast is headless. One paleontologist told me that many delicate bones comprise the skull, perhaps making it a savored part of anatomy among Mesozoic predators. So for nearly twenty years Carnegie's great giant remained headless, having fallen prey, it seemed, to finicky Jurassic eating habits.

We wanted to photograph Jack with the famous Carnegie Museum brontosaur, but the curator told us that the photograph would have to be taken after hours, when the visitors would not be disrupted.

It's nightmare time in the gallery of dinosaurs at the Carnegie Museum, as eerie shadows of the *T. rex* are cast on the walls and glass cases of skeletons. As we set up for our photograph, Jack walks around the

THE REAL BIG-GAME HUNTERS

famous dinosaur, smiling and admiring it like I imagine he did when he was a child.

As I join him on his anatomical tour around the brontosaur skeleton, Jack explains the mystery of its head. "What happened was that the original skeleton of *Apatosaurus* that Marsh found came from Morrison, Colorado, in 1877. An even better skeleton was found by his field crew in Como Bluff, Wyoming, in 1879 and was called *Brontosaurus*. Twenty years later, in 1903, Elmer Riggs showed that the two skeletons belonged to the same genus and the older name *Apatosaurus* should apply. This was the most complete sauropod that had been found up to that time, but it was missing its skull. When Marsh drew the first restoration of a sauropod back in 1883, he had to put a skull on it to make it look right. So he dotted in a skull

from a different animal, a *Camarasaurus* found at a different quarry four miles [6.4 k] away."

Marsh's understandable mistake set a trend in *Apatosaurus* headgear that was to last nearly a century. All around the country's museums *Apatosaurus* began sporting the now-fashionable heads of *Camarasaurus*, which were more sturdily constructed and in ready supply.

When researching old Carnegie Quarry records, Jack discovered that when the original Carnegie Quarry *Apatosaurus* was unearthed, it was found on top of a slightly smaller *Apatosaurus*, but neither had a skull. About two or three feet [61–91 cm] from the forefoot of the large *Apatosaurus* the diggers unearthed a small skull, about the size of a pony's, with little peglike teeth. At the time this was not considered suitable cranial equipment for an *Apatosaurus*, an animal with the

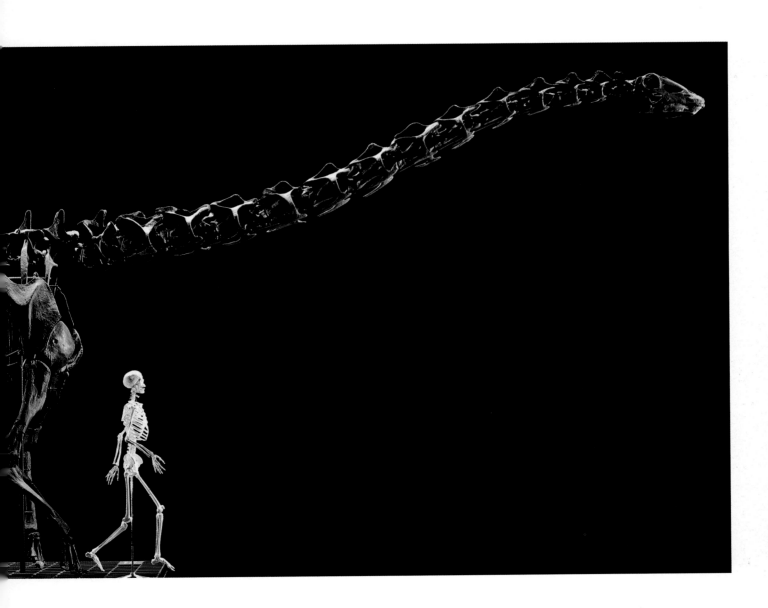

This *Diplodocus longus* from Wyoming, originally described by O. C. Marsh, is mounted in a state-of-the-art pose in Senckenberg Nature Museum in Frankfurt.

combined mass of five African bull elephants. A third skeleton, with only its back half intact, was found about ten feet [305 cm] to the side of the *Apatosaurus*. Identified as a *Diplodocus*, it was thought that the small head might have belonged to this creature, which had about a third the mass of an *Apatosaurus*. Since the Carnegie Museum was deluged with bones from the quarry and was running out of storage space, it gave away parts of this *Diplodocus*, minus the head, to a half dozen different museums, including a small one in Reading, Pennsylvania.

"When William Holland, the director of the Carnegie Museum, saw this skull," Jack explains, "he fitted it onto the first cervical [neck] vertebra of the large *Apatosaurus* and it seemed to fit perfectly. He wrote a paper in 1915 suggesting that it might belong to that skeleton, and he threatened to mount this skull on the Carnegie specimen. But Henry Fairfield Osborn, who was the great prominent vertebrate paleontologist at the time [he later became president of the American Museum of Natural History in New York], said he would make Holland the laughingstock of paleontology if he dared to mount it that way. So Holland didn't mount it, and he didn't publish his ideas, either. He left the skeleton without a head, and that was the way it was mounted in the Carnegie Museum until Holland died in 1932. Then the museum decided it ought to put a skull on the *Apatosaurus*, so it, too, like the American Museum of Natural History, put a *Camarasaurus* skull on it.

"In going through the correspondence between Earl Douglass [the discoverer of the quarry that became Dinosaur National Monument] and Holland, I determined that the large skull probably belonged with one of the two *Apatosaurus* skeletons," Jack tells me. "It was an accounting error. The numbers on the skulls were inadvertently switched. The size and description of the skull which Douglass gave Holland matches up perfectly with the larger skull. Dave Berman [the curator of fossil reptiles at the Carnegie Museum] and I published a paper in which we said we thought it probably belonged to it, and that was more or less accepted. Various museums have now changed their skulls and put this *Diplodocus*-like skull on their skeletons. But then I got a dividend many years later when I visited the little museum in Reading, Pennsylvania, which had parts of this so-called *Diplodocus* skeleton that had been found next to the two apatosaurs. When I looked at these bones, I just about jumped out of my skin, because they were not *Diplodocus* at all, but another *Apatosaurus*. So there wasn't even a *Diplodocus* anywhere near where the big skull was found!"

Even though Jack is a physics professor, and sauropods seem to defy the laws of physics, his interest in them concerns systematics—who begat whom. "Early sauropods are a big mystery," he explains. "They appear in the Lower Jurassic Period, sort of full-blown. They seem to be related to another group of animals called the prosauropods, which are something like them, but I think the relationship is not nearly as close as people originally thought. About the Lower Jurassic a rather large animal called *Vulcanodon* simply appeared in Zimbabwe, and that seems to be about the first one. But that was a big animal already, and what exactly it came from and what its direct ancestors are remain a big mystery."

Sauropods are seemingly design parodies—they have the biggest bodies of any land animal ever, and ridiculously long necks, also the longest of any animal. Some species such as *Mamenchisaurus* flaunt eight times more neck than a giraffe. Then to compound exaggeration, sauropods grew incredibly small heads, about the size of a five-gallon jug, through which it seems they would have to gorge themselves nonstop to support their massive bulk.

"They were eating machines," McIntosh marvels. "I think that's all there was to it."

The *Barosaurus* mount, erected in 1991 in the central hall of the American Museum of Natural History in New York, has been one of the most controversial displays in recent memory. This specimen also came from the Carnegie Quarry, but for some very complicated reasons various parts of its anatomy were lying in three different museums. The neck was lounging on a shelf

OPPOSITE: Jim Farlow uses a displacement theory developed by R. McNeill Alexander of the University of Leeds in England to calculate the weight of *Mamenchisaurus* at about twenty-three tons.

in Washington, D.C., at the Smithsonian, its body and limbs resided at the University of Utah, and a good chunk of the tail ended up at the Carnegie Museum. Then in the early 1990s Barnum Brown, paleontologist of the museum, who had no hand whatsoever in the *Barosaurus* excavation, orchestrated a trade in which he acquired this marvelous specimen for his collections.

In Barnum Brown's day, sauropods were thought to be aquatic, or at the very least, semiaquatic; and when they had to leave the water, some paleontologists thought they couldn't support their huge bulk on land for long.

After Brown's death, the *Barosaurus* was exhibited in part and only briefly, and lay unassembled in the museum's basement on Central Park West in New York. Recently, museum curators decided that the main entrance, always a great empty hall, could be used to better effect.

"There's a new administration here," Mark Norell, assistant museum curator tells me, "and there's been a real attempt to try to portray the museum in a different light, rather than just have dusty old bones around. The rotunda is a great space, it's a huge room. So we decided to do something exciting and provocative. Well, a lot of museums have big sauropods mounted and stuff, but we decided to do the *Barosaurus*, because we had the only good specimen ever collected. And we could put it in a really provocative pose, and when people would walk the stairs up into the gallery off Central Park West, they would be able to open the door and they'd just see this thing and they'd just go 'Wow.'"

When I first met Jack McIntosh, he was in Toronto, where a cast of the *Barosaurus* was being put together for the first time. As the leading expert on sauropod anatomy, he was brought in as the American Museum of Natural History's consultant to check the proper articulation of this controversial mount. It was being assembled in a parking lot outside a dinosaur factory whose ceiling couldn't accommodate the colossus. Workers from other businesses, breaking for lunch, stood transfixed by their cars, heads thrown back and mouths agape, as a large crane and workers on a giant scissors lift attached *Barosaurus*'s head to its neck some

ABOVE: Parts of this 140-million-year-old *Barosaurus* from Dinosaur National Park near Jensen, Utah, once resided simultaneously at three different museums, until dinosaur paleontologist and wheeler-dealer Barnum Brown finagled its relocation to New York City in the 1930s.
In 1991 the *Barosaurus* in the Teddy Roosevelt Hall reared skyward to protect its young from an attacking *Allosaurus*, a position that has inspired much scientific debate.
OPPOSITE: *Barosaurus* defends her young from attacking *Allosaurus*, by John Gurche, American Museum of Natural History, 1991.

fifty feet (15 m) off the ground. A large hawk circled above the head of the *Barosaurus* as if it were contemplating lunch. I asked Jack, as he gazed skyward, if he really thought a *Barosaurus* could rear up like this. He hesitated while searching for a diplomatic response. "If they could," he replied, not taking his eye off the dinosaur, "this is what they would look like."

In the display the *Barosaurus* was standing up to ward off an *Allosaurus* from making lunch of its baby. After the rotunda opened to the public, attendance shot up like an earthquake jolt on a Richter scale.

Nearly a year after *Barosaurus*'s installation and the media frenzy that its stance inspired, I ask Jack how he feels about the display.

"I'll just say this—I am not responsible for the pose of the *Barosaurus*, and as a matter of fact, I would have been chicken and would never have mounted it that way if it were my responsibility. However, I'll give you my opinion. There are three questions I'm asked. Did *Barosaurus* walk on its hind legs? The answer is certainly no. It was not bipedal. The second question is, Did *Barosaurus* and various *Diplodocus*-like animals get up on their hind legs to feed? If I were asked to guess, although it's very controversial, I would not be surprised if these particular sauropods could indeed get up to feed this way. One argument which interests me is this: The *Diplodocus* family, which includes *Apatosaurus* and *Barosaurus*, have the shortest forelegs of any of the sauropods. They have short trunks and long necks. Now why would these animals that have developed the longest neck, in order to help in feeding, have the shortest forelimbs? In other words, the neck was getting longer and the forelimbs were getting shorter. That would seem to be a contradiction in utility. However, if they were able to stand up to feed, then of course short forelimbs would be an advantage because there would be less hulk to pull in the air.

"The third question is," he says, "suppose that a predator came along and got after the baby: Could *Barosaurus* temporarily rear up? And there I think there's a much better chance that the answer is yes."

I ask him which was the largest mounted sauropod in the world.

"Oh, by far the big *Brachiosaurus* in Berlin. It's not the longest, but it's certainly the largest. The Field Museum in Chicago recently mounted a *Brachiosaurus* of similar size."

"And what about the longest mounted one?" I ask.

"The longest mounted one, I suppose, is the *Diplodocus* in Pittsburgh," says Jack. "The trouble there is, the curator has taken off some of the tail vertebrae and put them in storage because kids were stealing them."

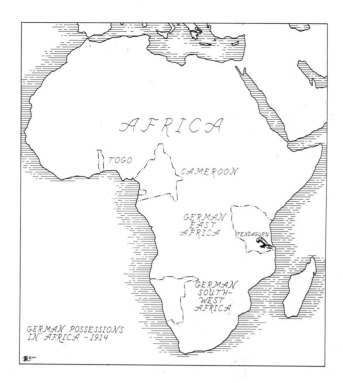

ABOVE: Tendagura Hill, Africa, where *Brachiosaurus* originally hailed.
BELOW: From 1909 to 1913, in excess of 250 tons of fossil material, including *Brachiosaurus*, was transported over four-day marches on the heads and backs of hundreds of native porters from Tendagura, then called German East Africa, to Lindi, a port town on the Indian Ocean more than fifty miles (80 k) away.
OPPOSITE: The largest mounted dinosaur in the world, *Brachiosaurus*, a 145-million-year-old vegetarian from Tendagura, now resides at the Natural History Museum of Humboldt University in Berlin.

Seismosaurus: The Glow-in-the-Dark Dinosaur

Deep within the top-secret Los Alamos National Laboratories would seem to be about the last place on earth you'd find scientists conducting dinosaur research. There some of the great scientific thinkers of our day have been involved in developing nuclear weapons. And now, strangely, they are analyzing bones of the longest gargantuan to walk the planet, *Seismosaurus*, "the earth-shaking lizard."

"When I was dinosaur curator at the Museum of Natural History in Albuquerque," Dave Gillette explains as I drive with him the hundred miles (161 k) from there to Los Alamos, "all the curators had an invitation . . . to come up and meet some of the scientific staff of Los Alamos. It was kind of a community service. And I think, *Oh, shoot, I don't need to go there. That's all high-tech stuff. I'm basically a Victorian-type scientist. I'm interested in anatomy and taxonomy and field work, and Los Alamos doesn't have any relation at all to what I'm doing.* But on a whim I jumped in the van. Rather than talk about all my accomplishments and achievements in paleontology, the way we usually do in seminars, I decided to talk about my problems. One was this problem in excavation that we have, never knowing for sure how much of a dinosaur lies buried in the ground. So I asked them if there was any instrumentation in their domain, secret or otherwise, that allows us to look into the ground, like having X-ray eyes. And the other question I asked is if there was any possibility of seeing ghosts of the soft parts of dinosaur or any other fossils, looking at chemical signatures as opposed to what we see visibly with bones. There were about a hundred scientists there, and both of those questions got these guys so excited that I was swamped with information after my talk. I couldn't believe the response. They clambered all over me with ideas on equipment they had for looking into the ground with remote sensing, and also on ideas for isolating proteins to look for genetic material which might be preserved in shales and sandstones. Several people in that audience took a leadership role organizing the Los Alamos

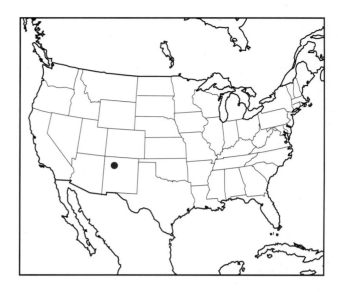

ABOVE: *Seismosaurus* site between Albuquerque and Los Alamos.
BELOW: Paleontologist Dave Gillette, who described *Seismosaurus*, "the earth-shaking lizard."

scientists, and we ended up doing primary experimental research: ground-based remote sensing in the field and chemical studies on the soft tissues."

The first accomplishment of Gillette and the Los Alamos crew was to extinguish a century-old myth about how buried bones become fossils.

"I always taught my students about the process of preservation of fossil bone—that it's preserved through this mysterious process of "molecule by molecule replacement" of the original bone until it is replaced by minerals. I'm sure you've seen that in books. That's become more or less a dogma. Well, that idea, as it turns out, was introduced in the 1860s in a casual paper by a paleontologist who really didn't know what he was talking about. That phrase was repeated over and over until it became truth. Paleontologists have hidden behind that explanation because it was a subject we never really cared about. But it always bothered me that when you look under a microscope at a thin-section of dinosaur bone, you can see crystal-clear contacts between the bone cells and the quartz that fills in the pore spaces. It always impressed me that we had such wonderful preservation of cell structure in bone, right down to the electron microscope level. It's as fresh-looking as modern bone. Except for the color, you can't tell them apart.

"The result of all that research was that we've demonstrated that the bone you see under a microscope—not the pore spaces that are filled with minerals from ground water, but the rest—is original bone. It contains at least a fraction of original organic material. We've isolated proteins from *Seismosaurus*, so there is a prospect of being able to assemble the amino acids into a sequence and determine genetic information. It may even be possible to isolate large sections of genetic code, so we could generate tissues, or even clone dinosaurs. I don't think that's unreasonable. But even if we can't do that, it will allow us to do chemical comparisons and determine if dinosaur bones contain

Seismosaurus, one of the longest dinosaurs ever discovered, is estimated to have been up to 150 feet (46 m) long, half the length of a football field.
Illustration by Pat Redman

extinct proteins or if they were identically qualitative to modern bone proteins, of which there are about four hundred in any given bone.

"The biochemical applications are really profound," he adds. "Proteins and organic compounds in bones almost certainly go through a predictable series of degradational events, which are controlled by temperature and pressure during burial. We may be able to reconstruct geologic history using the degradation products as a barometer that allows us to tell exactly how deep and how hot the bones got. That's an important application in exploration for hydrocarbons, oil, and gas. The applications just go forever."

Seismosaurus, discovered by hikers north of Albuquerque in 1979, is reportedly the world's longest dinosaur, reaching up to 150 feet (46 m). *Seismo* has been excavated for over seven years, and it will take perhaps another ten years to prepare and study it.

"There was ten feet [305 cm] of hard sandstone overburden, sometimes as hard as concrete," Dave explains. "It covered an area as big as a couple of swimming pools. So it was a huge excavation. We figured we moved over a million pounds of rock. We're maybe a fourth finished on preparation of the excavated material. We have around ten man-years of preparation left."

Preparation of *Seismo*, which progresses at about one square inch (6.452 sq cm) per day, was further

ABOVE: The *Seismosaurus* site in the upper part of the Morrison Formation, about 154 million years of age, was discovered by hikers looking for petroglyphs north of Albuquerque, New Mexico. It took about seven years to excavate the dinosaur because of the concrete-like consistency of the rock surrounding its bones. A skilled preparator working at about one square inch a day will need about ten years to complete the preparation.

RIGHT: Hot prospects from the Cold War—some of the more than one hundred scientists from the Los Alamos National Laboratories Paleo/Archeo Research Team who studied *Seismosaurus* in their free time.

OPPOSITE: Near the rib cage of *Seismosaurus*, Gillette's crew found about 240 stomach stones (gastroliths), enough to fill a ten-quart (10 liter) bucket. As with birds, Gillette thinks the stomach stones were used by *Seismosaurus* to pummel plants to aid digestion. While most of the stones were walnut-sized, Gillette speculates that this big one, as large as a shriveled-up grapefruit, may have actually choked and killed *Seismosaurus.*

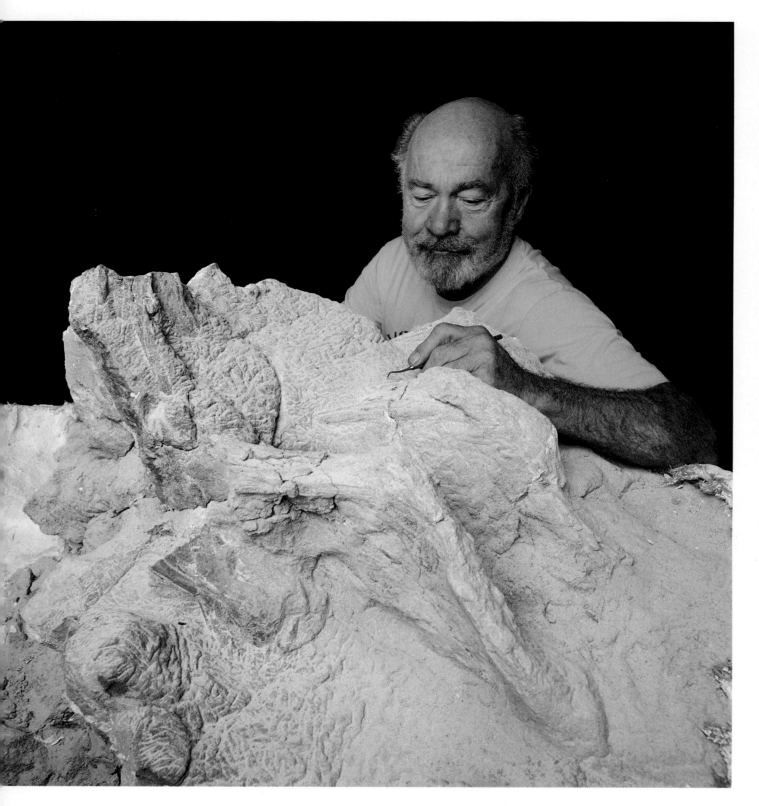

ABOVE AND OPPOSITE: The bones of *Seismosaurus* and the stone surrounding them were the same color, making preparation extremely difficult. Bones from the Morrison Formation are radio-active, approximately two hundred times more so than the rock encasing them. Gillette's team discovered that black light made the bone glow in the dark, expediting the preparation process.

slowed because the surrounding concrete-like sandstone is the same color as the bone. This problem was eliminated when Dave and the Los Alamos crew discovered that *Seismo* glows in the dark.

"That's a discovery we made quite by accident," Dave admits. "I made a bad joke at the *Seismosaurus* site on one field trip before I knew these Los Alamos folks very well. Some of the other scientists had been out there before and had taken Geiger counter measurements on the bone. It's common knowledge that the dinosaur bones in the Jurassic Morrison Formation are especially radioactive, but not at dangerous levels. I was leading a field trip out to the site for about fifty new people from Los Alamos, mainly physicists and chemists, people who really didn't know any geology. They were excited and attentive and very intense, [and] I just casually mentioned that the bones had a pretty high uranium content but were safe to work around. I didn't know what to say after that, so there was a long pause and I said, 'Except we glow at night.' And there was dead silence. Nobody thought that was funny at all. These are the guys who build nuclear weapons. It was one of the worst faux pas of my life. And then one woman who hadn't said a word all day but was right in front all through the tour of the site said maybe the bones glowed under a black light. Everybody started into conversation, thinking that was a brilliant idea. So we took a fragment of bone back to Los Alamos and went into one of the labs where one of the guys had a black light. But the lab wasn't dark enough, so about a dozen of us went into a men's bathroom without any windows. We were clambering all over each other, and when somebody flipped off the overhead light and flipped the switch for the black light—the bone glowed. It was sort of like rediscovering the wheel."

Later Gillette and I drive up to a high-security checkpoint of Los Alamos Labs. After we show the appropriate documents, the guard waves us through. As we approach the barrackslike labs, about ten scientists greet us, all wearing red T-shirts with "Seismosaurus" emblazoned on the front. The Cold War has some hot new prospects for dinosaur researchers, I think.

DINOSAUR JIM: HUNTER OF MEGAGIANTS

On a mountaintop outside Provo, Utah, legendary dinosaur hunter Jim Jensen, now retired, is sitting on the tailgate of a pickup truck watching a crane raise the foreleg of *Ultrasaurus*, probably the world's largest known dinosaur.

Jensen's career was punctuated by spectacular discoveries. Among the most renowned are *Supersaurus* and *Ultrasaurus*, vegetarian megagiants who were two of the biggest creatures ever to walk the earth. Their discovery in the 1970s in the Dry Mesa Quarry, Colorado, earned Jensen an international reputation and the nickname "Dinosaur Jim."

As Dinosaur Jim reminisces about a career in dinosaurs, he gives the impression of being a cowboy who stumbled into science.

"A mammal hunter can come back from a summer's worth of finds in a shoe box and feel he's eaten a

BELOW: Provo, Utah, site of Brigham Young University
OPPOSITE: From Dry Mesa Quarry in Colorado, Dinosaur Jim Jensen has excavated the shoulder blade of an animal he calls *Ultrasaurus*, perhaps the largest animal to ever walk the earth. He stands with the extrapolated cast of its foreleg hung from a crane.

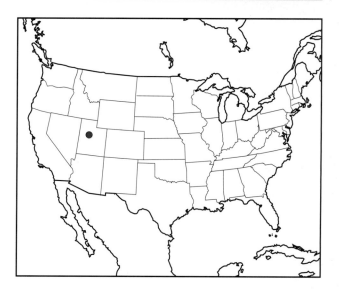

pretty good hog," he says with an eye trained on the *Ultrasaurus* leg towering above us. "But a dinosaur guy can't put anything he finds in a shoe box. He'll fill up all the museum's storage space in a summer."

When Jim, who is a Mormon, retired from Brigham Young University, where he was curator of the Earth Sciences Museum, he left a hundred tons of dinosaur material to be worked on, about a hundred years of preparation in the laboratory. After Jensen ran out of storage space at the museum, the university let him store his legacy in the only place big enough to contain it—under the football stadium.

"I know there are at least two dozen new dinosaurs in there," he says, "all previously unknown. Since I left there, the preparator has been opening up blocks that I collected and discovering marvelous things."

In 1972, with the help of rock hounds [amateur private collectors] he had enlisted to locate new sites, Jim discovered what became one of the most productive dinosaur quarries in the world. The quarry at Dry Mesa, Colorado, is still being worked by his successors at Brigham Young.

"The variety of animals at Dry Mesa was fascinating," Dinosaur Jim explains. "Every day we worked in

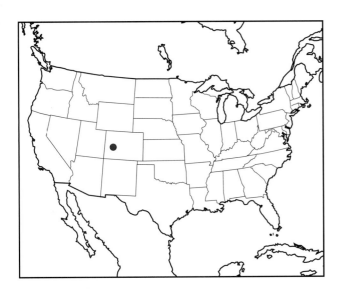

ABOVE: Dry Mesa Quarry, near Delta, Colorado
OPPOSITE: The Earth Sciences Museum's storage space having run out, Jensen's legacy, over one hundred tons of unprepared material still wrapped in plaster jackets, lies in storage below the football stadium bleachers at Brigham Young University.

the quarry, I'd see a bone that I did well to figure out what part of the animal it was from, let alone to whom it belonged. Every dinosaur we dragged out and made the acquaintance of was a new creature. It was exciting working there, but the wind blows sand in one ear and out the other, and gnats crawl up into your hair when you got your hands in plaster. Let me tell you, you have to be dedicated to paleontology. The people I took there that weren't didn't last very long."

John and I made a pilgrimage to Dry Mesa Quarry, a few hours' drive from the nearest town, bearing a case of beer as an offering to the crew, not realizing that everyone in the crew was Mormon and drinking was forbidden. One harried gentleman who became our guide through the site had taken refuge in the heat and hard labor to get away from his three wives and offspring, which numbered deep into double digits. As he gave us a tour of the gigantic site, he admitted that nobody really knew how all those animals came to be buried there.

"I don't know what killed them," Jim says. "I suspect some died of old age, but some were killed by carnivores. I have bones from there with big teeth marks where they were gnawed on by meat-eaters. In fact, one dinosaur I found and named is the strongest meat-eating dinosaur ever found. It had an extremely powerful forearm with claws a foot and a half long, and it had a very powerful skull. The whole skeleton was very powerful—it was a sauropod killer. My first knowledge of its existence came when I found a sauropod skeleton that had been partially eaten by a huge carnivore, larger than anything we knew from the Jurassic. I looked for four years in quarries before I found the first bones of that big carnivore. I named it *Torvosaurus tanneri*. While *Tyrannosaurus* has little tiny forearms, very ineffectual, can't even scratch its ear, *Torvosaurus* had a very powerful forearm, which it needed to rip into the thick hide of sauropods."

Two of the biggest critters to come out of Dry Mesa, sporting the biggest frames of any animals in the world, were Jim's famed *Ultrasaurus* and *Supersaurus*, sauropods known from only a few isolated bones.

"The *Ultrasaurus* is supposed to be a hundred feet [30 m] long," Jim said. "I didn't come up with that fig-

Ultrasaurus, an animal known from a single bone, its shoulder blade, may have reached heights of up to fifty feet (15.24 m).
Illustration by Pat Redman

find all these exciting things. But my sympathies are not with the Ph.D.s. It took me forty years to graduate from high school. Well, when they hired me here at BYU, the dean told me, 'We'd like you to work on your degree.' And I thought, *Yeah, my high school degree.* So I enrolled myself in a night school and became the only faculty member working my way through high school. Forty years later my old high school asked me to be the baccalaureate speaker and I said, 'Would it be possible for me to graduate?' They got a diploma ready and I marched with the graduates and received my diploma. But then, when you're in paleontology you realize things move by slow. Large increments of time never bothered me. BYU gave me an honorary Ph.D., which was very useful for conducting international business with other museums when you're in a phone-booth-sized operation."

After our photo shoot, the crane starts to lower the leg of *Ultrasaurus*, and Dinosaur Jim is ready to go. He becomes philosophical about the future.

"We're all headed for the exit," he says as he adjusts his trademark safari hat. "When you get to be my age, you start thinking about that big boneyard in the sky. I sometimes dream that there are dinosaurs in hell. I hope there is, so I'll have something to do—if there isn't, it's going to be hell."

ure—that was estimated by some armchair specialist. I wouldn't hazard a guess how long they were until I found three-quarters of them. They must have stood very tall, because I found a neck vertebra that was nearly four and a half feet [137 cm] long, just one vertebra. If *Ultrasaurus* held its neck out giraffe-style, it would probably stand around fifty feet [15 m] tall. The media squeezed that out of me, fifty feet [15 m] tall or more. But exactly how big it was I don't know. I don't know if we'll ever find enough to establish that. We can go by *Brachiosaurus* and extrapolate and get an *idea* of size, but it's not fact."

Although Dinosaur Jim has inspired numerous Ph.D.s, his own career lacks some of the academic luster of his colleagues and disciples.

"I'm not a highly trained scientist," he explains. "I am an adventurer who happened to wander into paleontology. I had a marvelous time, got to travel and

190 MILLION SOMETH'N'

"I got started in 1930 when jobs were real tough, and I got a job here for fifty cents an hour. And that rubbed my nose into paleontology and I couldn't get out. I tell you when you get into this field there's just no limit to where you can go in any direction and it's so challenging. It's so inspirational. Once you get in here and get to working in it, you lose yourself."

Octogenarian bone hunter Samuel Welles is Museum Scientist Emeritus at the University of Califor-

OPPOSITE: Octogenarian bone hunter Sam Welles, researcher at the University of California at Berkeley, with a cast of *Dilophosaurus*, the "double-crested reptile," a Jurassic-aged carnivorous dinosaur found on the Navajo Indian Reservation in Arizona.

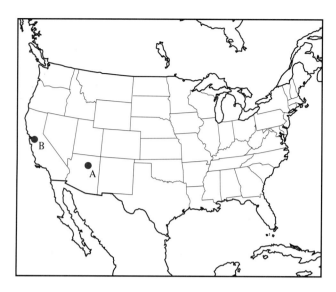

ABOVE: Site of the *Dilophosaurus* dig on the Navajo Reservation near Tuba City, Arizona (A), and the California Academy of Sciences in San Francisco (B).
BELOW: Illustration of *Dilophosaurus* by Shannon Shea

nia at Berkeley. For over sixty years he has been best known in the fossil world for his work with the sea monsters called plesiosaurs, those Loch Ness–like monsters that seem to have been about as common as trout in the Jurassic and Cretaceous, but that's another story. Things that lived in the sea weren't dinosaurs, but Sam found a couple of strange land-based monsters that have kept him busy for the last fifty years.

"I happened to get into this dinosaur *Dilophosaurus* that a Navajo Indian found and I collected. That was near Tuba City, north of Flagstaff in the Little Colorado Valley, which is in the Kayenta Formation, which is Jurassic. I hired the Navajo fellow who found it, to help him out. We didn't have a budget for a reward or anything like that. He came in and he worked with pick and shovel for a couple of days. And then when you are collecting a good specimen like that, when you get close to the bone, you don't use a pick and shovel, you use smaller tools, a whisk broom

and then, finally, fine brushes with just a few hairs. We started to do that and he disappeared. We thought he quit. He came back two days later with two squaws, and he sat them down and started to tell them to go to work. I said, 'I'm sorry, we can't do that. We've got to do this ourselves.' He said, 'That's not man's work, that's squaw's work.' Maybe it was, but as a paleontologist, you do them both. It was beneath his dignity and so I had to let him go. We got that skeleton in 1942 and then went back many years later (1964) and found another skeleton of *Dilophosaurus* in the same level about a quarter of a mile away. And that one has yet to be described."

Until the second specimen (which was much better preserved) was prepared, Sam hadn't recognized the unique headgear that would give this animal its name. *Dilophosaurus*, which means "two-crested reptile," had two thin bony crests that adorned the top of its head.

"Until I found the second specimen I didn't recognize the crest," Sam laughs. "I didn't expect to see any crest, it's sort of like finding wings on a worm—you don't expect them. I can't tell you what they were for. They could have been for heat regulation or they could have been simply for display. They certainly were not useful in fighting, because they were paper thin, very delicate. They spread apart a little like a V, so there was an open notch down the center of the top of the head between the two crests. *Dilophosaurus* was a dinosaur standing about eight or nine feet [244–274 cm] high, maybe something a little less, about the weight of a pony, I would think. With very sharp front claws, very sharp hind claws. Walked on his hind legs, of course, like all of the meat-eating dinosaurs. One thing about *Dilophosaurus* which we are beginning to find in other dinosaurs is that the hand is flexible like a human hand, the thumb bends in. That's something we hadn't appreciated before. These animals were not primitive. They were advanced, they were active and they could run and kill with their teeth, their big sharp claws on either their front or hind legs. They ate anything they could find. I assume they ate anything that was dead as well as anything they could catch and kill."

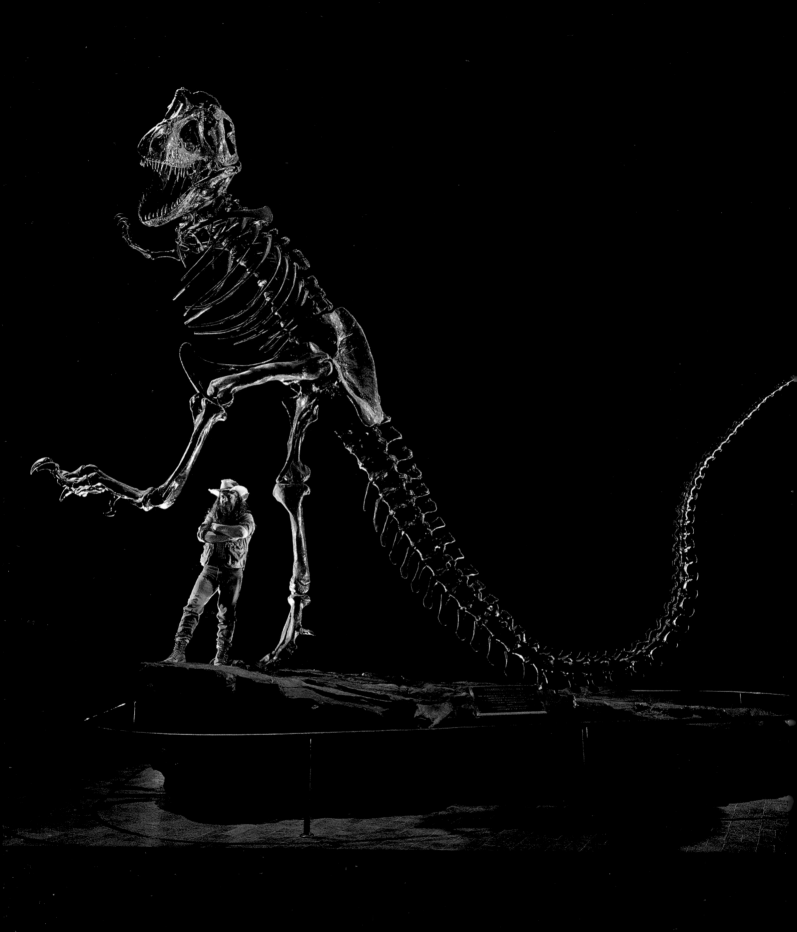

WARM-BLOODED THEORIES

DEEP INTO THE REAL JURASSIC PARK WITH PALEONTOLOGY'S HOT-BLOODED DESPERADO BOB BAKKER

Bob's Datsun pickup was clogged, as usual, with paleontological supplies, and on his dashboard the deteriorating skull of "Rocky" the raccoon was ripening nicely into an eerie mascot to another world. And another world was where we were headed. The annual Gem and Mineral Show in Denver is a strange affair, and Bob, who lives up the road in Boulder, thought it was about time he saw one.

For one week each fall, rock hounds and private fossil collectors from around the world take over entire hotels in downtown Denver and transform conventional suites into fossil shops. Black velvet is hung over the windows so fossils can be dramatically lit with jewelers' lamps. Visitors are permitted to wander into the open hotel rooms stuffed with bones and skulls neatly arranged on the beds and night tables. Bizarre extinct anthropods called trilobites fill the dresser drawers, and dinosaur bones rest on couch cushions in the bathtubs. It is Open House in the Holiday Inn from Hell, and among these denizens of the fossil world, the paleontological

ABOVE: In 1974 Bob sent shock waves through the scientific community by declaring that dinosaurs are not extinct but live on through their direct descendants, the birds, which, he argued, belong in the dinosaur family tree.

OPPOSITE: Known worldwide for his heretical ideas on dinosaur behavior, Bob first shocked the scientific community in the late 1960s by arguing that dinosaurs were active and warm-blooded like mammals. As his warm-blooded theory gathered the widespread support of the scientific community, museums around the world responded by mounting their dinosaurs in more active poses. Bakker granted this *Tyrannosaurus*, cast at the Denver Museum of Natural History, a pose for which it has been called "the chorus-line *T. rex*" for its Rockette-like aspect.

outlaw Bob Bakker is a celebrity. Word of his arrival spreads quickly down the hall, and giddy fans cluster in doorways trying to look inconspicuous as they glimpse at the man who almost single-handedly restored the reputation of the dinosaurs. Fans approach him to autograph their T-shirts. One collector steps out of a rock shop/hotel room with a sign outside that says ART BY GOD and asks Bob to identify a strange bit of bone. After Bob does so, the collector walks away with a serene smile like a believer who has just seen the pope.

Up until the early sixties the Dinosaur Kingdom was suffering from a lot of bad press. Dinosaurs were routinely thought of as cold-blooded, slow, and stupid swamp dwellers, evolutionary delinquents whose time for the genetic scrap heap had come. The very word "dinosaur" had become a metaphor for large historical mistakes in our own culture. But Bob Bakker began changing all that. Beginning with his thesis in 1966 at Yale University, he started championing the heretical concept that dinosaurs are warm-blooded like mammals. He is an excellent anatomist and argued that bulky grazers like *Triceratops* had the anatomical equipment to gallop. He cited evidence from trackways that suggested that animals like the brontosaurs were not swamp-dwellers but migrated on arid plains, like elephants, with their young at the center of the herd for protection. His warm-blooded theories captured the attention of the world, and eventually nearly every major museum responded by displaying their dinosaur specimens in more Bakkerian positions. Before Bakker, *T. rex* was nearly always depicted as a lethargic but vicious kangaroo dragging his limp tail along the ground. When he was invited to restore the *Tyrannosaurus* skeleton in the front hall of the Denver Natural History Museum, Bob gave it the poise of a ballet dancer about to leap. He not only gave dinosaurs back their reputation, he claimed that they are alive and well in the form of birds, their direct living descendants.

Dinosauria's main heretic has a bachelor's degree from Yale, a doctorate from Harvard, and the frontier wisdom gleaned from thirty years of digging dinosaurs out West. Bob sports a broken-down cowboy hat that should have been thrown away ten field seasons ago. It has become as much of a personal trademark as his scraggly beard. Bakker is delightfully irreverent and always humorous. He speaks in colorful sound bites that reporters love. He is sought after for lectures, interviews, and dinosaur TV specials.

But not everybody likes Bob. Some members of academia regard him with varying measures of awe and contempt. He is outspoken in his challenge of colleagues' outdated notions. His published challenge to dinosaur dogma, called, appropriately, *Dinosaur Heresies*, helps finance his field season and along with his lucrative speaking engagements ($10,000 per lecture) unleashes him from the academic system he professes to abhor.

Scientists publish their peer-reviewed results in accepted scientific journals. The process of publishing can take years if not a good part of one's career, and the public may never hear much of a scientist's life's work. Bob short-circuited that system. He publishes his own scientific newsletter and has a substantial network of media friends who eagerly await his latest radical theory. Bob's ideas can enter the public's con-

sciousness in a day. Some of Bakker's fellow scientists claim that his ideas are more suitable for science fiction, and indeed Michael Crichton, author of *Jurassic Park*, credits Bakker's theories as the book's inspiration.

Whatever his colleagues may think of him, it is roundly agreed that Bob Bakker has done more to put dinosaurs on the fast track of scientific study than any other human. It is impossible to talk to a dinosaur paleontologist today without eventually referring to the ideas of Bob Bakker, as they are all, consciously or not, trying to refute or support "Bakkerian theories." Trying to imagine modern dinosaur paleontology without Bob Bakker is like trying to imagine the sixties without rock and roll.

Bakker was out on the lecture circuit, and after leaving messages on his answering machine for several weeks, I received a cryptic fax drawing of a meat-eater slashing a herbivore followed by a message that said . . .

"Rock River, Longhorn Lodge, Wyoming, This weekend, dig dinos, bring herring, save bones, win prizes."

I tried to reach him again by phone to confirm, but I got his message advising me, "You have reached Sid the boa constrictor, and I can't come to the phone right now because I've just eaten a goat." I searched the map to find Rock River, bought a plane ticket, and headed out to the fossil fields of Como Bluff, famous for being the center of what was called "the Great Bone Wars" or

THE GREAT DINOSAUR RUSH
(AND THE FINE ART OF RUSTLING DINOSAURS)

In 1877 the American frontier was being pushed farther west by railroad crews who were putting a new line through what was still a war zone. Custer had been defeated at the Battle of Little Big Horn the previous year, so the West was far from won. The railroad companies hired marksmen to shoot antelope and buffalo to feed its workers fresh meat. One of these sharpshooters for the Union Pacific was William Harlow Reed, who was carrying the flanks of an antelope and his Sharps 50mm rifle over a hill near the newly built Como Station when he noticed some large bones

ABOVE: Arthur Lakes's painting of Como Bluff during the 1879 field season.
BELOW: Como Bluff, Wyoming (A), and Boulder, Colorado (B).

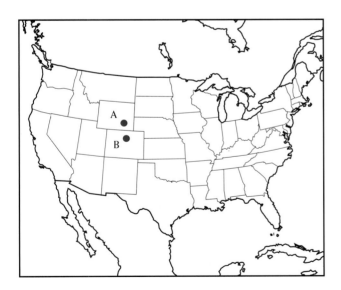

coming out of the earth. They were like nothing he had seen before. It was a *Camarasaurus*, as common 140 million years ago as the antelope he was holding, but weighing in at some twenty tons and measuring fifty feet (15 m) long. One animal would have provided dinner for every settler in the Wyoming territory.

Reed and station manager Bill Carlin contacted Othniel Charles Marsh, the Yale professor who was well known among railroad workers for his bizarre interest in extinct animal remains. Marsh had traveled through the area a few years before with his field crew and another sharpshooter, Buffalo Bill Cody. The Yale professor had shipped fossils back east by the boxcar load to furnish his newly created Yale College Museum.

Marsh had another, darker motivation to find dinosaurs. He was also aggressively trying to discover and name new specimens before his archenemy, Edward Drinker Cope, could do the same.

Marsh secretly sent the leader of his field crews from Colorado to investigate. With several miles of rich fossil fields yielding dozens of nearly complete specimens, Como Bluff quickly became one of the most important dinosaur sites in the world. This was the Jurassic mother lode. Never before had so many different, nearly complete dinosaurs been found. Over the next few decades the Great Dinosaur Rush furnished many of the still-existing museum displays.

I arrive at Rock River, a town with tumbleweeds blowing down Main Street. Bakker hasn't arrived yet, so after checking into the hotel, I play pool down at the only bar in town with the locals, who all work at the railroad.

At the site the next morning, the hills of Como Bluff are naked of vegetation, perfect for looking for fossils. Since bone hunters have been pulling dinosaur bones out of this place for well over a century, the mother lode may have been taken, but there are still many more to be discovered.

"There's a hell of an extinction here," Bob says, unpacking supplies from his pickup. "It's the Jurassic-Cretaceous twilight zone. That sloping, grassy plain is the very bottom of the formation. The whole thing is about three hundred feet [91 m] thick. It's the last

five million years of the Jurassic Period. Most of the bones we have are disarticulated. They're not 'mediagenic,' as we say. They're very delicate. The reason for working here is that we can walk through the Jurassic-Cretaceous boundary in a highly fossil-rich environment. The modus operandi for this extinction is the same as in the Cretaceous extinction. Whole families of sauropods, things like *Diplodocus* or *Brontosaurus*, *Camarasaurus*, *Stegosaurus*, *Allosaurus*, *Ceratosaurus*, *Megalosaurus*, which were very, very common, go extinct or become extremely rare. These are long-lived families which go back ten, twenty million years. Just gone. It's a spectacular extinction. It's not a total dinosaur extinction, but the style of extinction is exactly like the K-T boundary. It's absolutely fascinating. Carry my toolbox?"

Bob and his wife, Constance, use garage-sale ski poles as walking sticks as they maneuver their way up to the quarry a few hundred feet up the side of a steep slope. His field crew is already at work.

"It's easy to find bone in the Morrison Formation," he says as we hike to the site. "Used to be lots of quarries here. The bone is hard, full of natural glass, silica. But nearly all the good quarries are in the same level—in the upper one-quarter and more or less the same fauna. Previous collectors were going for bone that was harder and more articulated. Takes two, three, four, ten times longer per bone to work delicate stuff than typical hard bone."

"What have you found that's new?" I ask.

"Virtually everything. Because the extinctions were so rapid, if you go up fifty feet [15 m], you're losing species, genera, and families. And the fauna's much richer here in the lower half than in the upper half. We have a whole bunch of families here that don't occur higher up: the cetiosaurs, or short-necked sauropods; megalosaurs, very primitive meat-eaters; brachiosaurs, the first African-style brachiosaurs. Scientists need to think more like historians. There are repeated patterns over one hundred sixty million . . . Well, there are

OPPOSITE: Billed as the world's oldest house, a fossil shop near Bone Cabin Quarry is constructed of Jurassic dinosaur bones about 145 million years old.

repeated patterns over three hundred million years. There's different tempos, but there's a cycle.

"We'll never understand the terminal Cretaceous extinctions until we look at all extinction. We'll have to take a serial killer approach to solving it. It's not one crime of passion in the ecosystem that killed off the dinosaurs. It was this serial killer who operated in exactly the same way again and again and again and again.

"You're in the end of the Jurassic. Who's dying off? Little animals? No. Turtles and crocs? No. Snails? No. Fish? No. Little dinosaurs? No. Mammals? No. Just the biggest, most dominant dinosaurs are dying out. And who is surviving? Anything that's little. Anything that lives in fresh water. And anything that was rare. We have ankylosaurs—the oldest ankylosaur from North America—one specimen. It's outnumbered by stegosaurs a hundred to one. In the Cretaceous, the earliest Cretaceous, ankylosaurs suddenly get real common. It's like the extinction released these rare groups."

"There's some speculation," I say, quoting a recent extinction theory, "that extinctions are happening on a fairly regular basis of about twenty-two, twenty-five million years."

Bakker pounces quickly and emphatically. "Man, that's garbage. It's not that regular. Some of them are spaced five million years apart. Some twenty. And another key point is, there are more extinctions on land than in the ocean. Sometimes there are extinctions in the ocean at the same time as on land. And the Ice Age extinctions, there's no extinction in the ocean. Who did we lose in the Ice Age? Chipmunks? No. Frogs? No. Turtles? No. Snails? No. We lost big, common things, mammoths, mastodons, saber-toothed cats. Now that's interesting."

"So are you saying that every extinction that occurred might have had the same agent?"

"There has to be one agent. All of the extinctions look alike."

"If you had to guess, what do you think it is?"

"Well, you don't want to guess. You want to eliminate some things. It can't be climate. Big animals were robust in terms of climate change. It's very hard to kill an elephant by starving it to death, or depriving it of water. It will just move. And it can survive a long fast. Yet a chipmunk you can starve in a couple days. It can't be because the world gets colder. Or hotter. Or drier. Or wetter. Because that would kill the small animals first."

"Comet?" I suggest, offering the most recently accepted smoking-gun theory.

"Well, that wouldn't work, either. If you suddenly chill or heat up the earth with a celestial impact, who's going to die first? A tropical frog this big?" He holds his thumb and forefinger an inch apart. "Or something the size of two elephants put together? The little frog will die first because they can't stand a sudden change in temperature one way or the other. Or sudden pollution in the water. Tropical frogs today are extremely sensitive to acidification of their water ecosystem. Frogs never suffer mass extinction. They never have. Never. They're indestructible. Throughout evolutionary history there are some groups, like frogs, which you can't get rid of. You just get more of them. Right now we have more families of frogs than we have ever had in the history of life. Hardly any frog families have gone extinct in their whole damn history, and those facts have been ignored roundly by everyone who believes in the meteorite theory."

"If it's not climate," I ask, "what else could it be?"

"That eliminates nearly all hypotheses," he says. "It has to be something that only big, active, common animals on land do. Or happens to them. And it may or may not coincide with extinctions in the ocean. That's a real red herring, to look for one cause to exterminate animals in both the ocean and on the land."

We arrive at the site and he introduces me to his field crew, Jim and Louise, two volunteers who jump out of their pits to shake hands. They go back to work. Constance finds the rare shade of a bush and settles down to read a book.

"This is one of about twenty sites we have," Bob explains. "The purpose is not to get any one particular species or one fauna, but to get a sample of a whole changing ecosystem of the Morrison. We've been quite successful in finding many sites. When we started working, there were only two levels that were well known and now there's about twenty-four. About ninety percent of the quarries are above here. Of course that's where the bones are really hard, and easy to dig. If you are just starting out, it makes sense to get

after the bones that are easier. There were very hard silicified bones. Chunks of bones lying near the surface. You still occasionally find that stuff, but it tends to get picked up by rock hounds now and destroyed."

I ask Bob to describe what these hills looked like during the Jurassic.

"It used to be thought that this was all a swamp, like New Jersey. That turns out to be totally wrong. There are many, many different environments here. And most of them had severe dry seasons. About a hundred thirty-eight, pushing a hundred forty million years ago, end of the Jurassic, you would have seen thick gallery forests along the stream lines. A more open conifer forest in between. A lot of bare meadows where everything died back during the dry season. During the dry season, you might not have seen any dinosaurs at all. Or they migrated hundreds of miles away. But they would probably follow the rains like animals do in East Africa today. Maybe they came just before the rains. Or with the rains. That's when you see the herds moving in. Imagine a flat interior of the Great Rift Valley in Africa. That's what this was like. Long dry season, occasional floods, salt ponds, salt lakes, ephemeral lakes."

"How did all the bones get here?"

"You can go for ten thousand years without any significant rain. What was happening was quite episodic. You'll have a thick layer of gooey mud laid down in the wet season. Maybe six feet [2 m], maybe ten feet [3 m], at one time which will cover the bones of everything that died in the previous couple years. It covers and mixes the specimens. And then nothing for years and years and years. Ten years, maybe even a thousand years, there's very little deposition. You might go tens of thousands of years without another major flood. There was a whole series of rivers that had exceptionally heavy discharge in the mountains in Utah. And a very heavy fall of volcanic ash which was full of silica. And the silica is very soluble in alkaline water. The East African rift valleys today are often full of silica. The silica gel, that liquid glass, will penetrate a bone and then just freeze up. You've got a glassed bone. Really hard. Easy to dig. So you have the combination of high discharge of water, lots of streams choked with volcanic ash making this syrupy mud. Things will grow in that mud, changing its color, and

Bob uses a ski pole as a walking stick as he
explores the Jurassic badlands of Como Bluff with Prance,
his canine companion.

it will be pretty damn dry. Maybe in the worst part of a drought their remains were chewed up by carnivores. We have all these shed teeth. And then you'll have another flood.

"The animals that were buried in this flood are nearly all terrestrial. There are no fish, no snails, no clams, no crocodiles, no turtles. We have dinosaurs preserved in lake beds, but the lakes must have been toxic, because fish are incredibly rare and fish-eaters like turtles and crocodiles are rare except in certain localized little beds. And the turtles and crocs are small. Aquatic,

all right, but they're small. So the lack of fish, the rarity of crocs and turtles except in local little pockets, the fact that they are small, all indicate that, on average, there wasn't a lot of standing water. The second evidence is the chemistry of these sediments. A lot of them are full of salts that accumulate in a salt lake or a salt pond, so the lakes were inhospitable to fish."

Buried within this relatively small quarry, he tells me, there are two megalosaurs that are three-and-a-half-ton *T. rex*–like predators, one smaller predator, an allosaur, a stegosaur, two small adult sauropods, and a baby *Diplodocus* as big as an elephant.

I'm trying hard to imagine a flat African rift valley on this hillside in Wyoming, but in an area that was supposed to be so dry I wonder how the animals got so big.

"In the dry season, the only thing available to eat," Bob explains, "would be the very tops of the tallest trees that were still sucking up water from low in the water table. To get at the very tops of the tallest trees, not only do you have to be very big, you have to have a long neck. And sauropod dinosaurs are only common in diverse, semiarid environments."

I ask him if he thinks the *Barosaurus* mount at the American Museum is accurate.

"Oh, that's the original position as described by Marsh in 1880. And it's absolutely correct. Also discussed by Riggs in 1904. That's the way they work. The center of gravity is right in front of the hips. If you go like this . . ." He puts his forefinger on my chest and gives me a slight nudge. "If you just push them in the chest, they automatically straighten up. People say, 'Oh, how did they get blood up to the brain?' Well, the brain's only the size of a small green pepper. No problem at all to squeeze a few neck muscles and get some blood up there. Why do you think the head is so small in sauropods?

"The first complete skeleton found at Como Bluff was a sauropod with a completely intact braincase, really small for its size. The myth that dinosaurs were

Bob Bakker and crew excavating Jurassic dinosaurs at one of his Como Bluff, Wyoming, quarries. (In the center is one of the two Jims.)

ABOVE: A fleet-footed *Daspletosaurus* gets ready to make a casualty of a less speedy *Struthiomimus*.
Illustration by Bob Bakker

OPPOSITE: A brontosaur looms over other Breakfast Bench fauna in Latest Jurassic Wyoming while a pterodactyl removes its skin parasites. A hapless ornithopod dinosaur, *Drinker nisti,* is munched by a croc on the shoreline while in the water another *Drinker* is menaced by a lungfish and the turtle, *Uluops uluops,* swims by unconcerned. In the dead tree a multituberculate mammal, *Zofiabaatar pulcher,* protests the approach of *Foxraptor atrox,* a mammal.
Illustration by Pat Redman after drawing by Bob Bakker

pea-brained began with that specimen. Marsh called it *Morosaurus,* which literally means 'the moronic lizard.' "

"But that huge mass must have made life difficult."

"It's cheaper to be big," he says. "That's why elephants survive in droughts. Because they can move. They're selective about what they're going to eat. They can move hundreds of kilometers. If you're a dik-dik (one of the smallest antelope), being very small means you cannot migrate or escape a drought. You're restricted to pretty dense undergrowth, as are small plant-eaters. If you're big you have great freedom. Now you want the top of a tree, you either reach it or knock it over. The correlation between a dry habitat and big, big plant-eater is very close. It still works today. The white rhino is the biggest rhino and it's also the most dry-adapted rhino.

"The biggest Marsh quarry that existed is about a mile from here. That was the Stegosaur Quarry. And that always puzzled me, because *Stegosaurus* is usually real rare. You get one per twenty brontosaurs. Until I started looking at all the environments and found that every place that was rich in stegosaurs was either in a large riverbed or near one. So there are two dinosaur communities here: the sauropod community, which moves in and out with the dry season, and the stegosaurs, the animals that stayed close to permanent water, the river courses and lakes. That works in the Cretaceous, too.

"Look at that," he says, pulling back a blue tarp. "Isn't that beautiful. Isn't that just pretty bone?"

"They don't make 'em like that anymore," I say.

He sits cross-legged on the ground and begins brushing away dirt covering a large bone. He tells me it's an ilium of a megalosaur, part of the hipbone, on this creature about two-by-three feet (61-by-91 cm) and the biggest bone in the body. Bone in the Morrison Formation, for some mysterious reason, is very radioactive.

"The uranium was not there when these bones were deposited. They picked it up later. I don't know exactly why. Maybe simply because they're porous and the fluids with the uranium salts percolate through the bone. A little bit of dinosaur bone has even been ground up for ore in Wyoming and Utah in the 1950s. Not a lot. But a lot of the bone is hot enough so you wouldn't want to put it in your pants pocket and carry it around for a day."

I have a list of stock questions that I try to ask nearly every paleontologist. I ask him, "Why study dinosaurs?"

He stops digging for a moment and seems irritated, maybe a little angry, and for a moment I wish I hadn't asked him.

"Why learn how to play the bassoon? Why do people pay you to take pictures of mountains? Why? Because they're beautiful. Because they're interesting. Because they're riveting. They take your mind and they stop it.

"Dinosaurs are Nature's special effects. They're real monsters. And they're history. They are not just a mindless parade of prehistoric circus.

"It's a complicated history, yet that's why we're digging here. I mean, we don't get individually pretty specimens from here. But the pieces we're getting . . . It's just such a fascinating, complicated historical story, crisis/extinction. It really makes sense. There are regularities in evolution that are startling. Evolution's a spooky thing. Why is it that dinosaurs evolved fast, so

that every million years or so you get a new species, but turtles and crocodiles evolve slow?

"There is a richness of dinosaurology. Each time slice has its own animals interacting with each other and doing neat kinds of horrible things to each other. And those interactions, that warfare, ecological warfare, shapes the next layer, which shapes the next layer. This is not just a monster parade. These monsters have personalities. They're killing, they're eating, they're breeding, taking care of their young. And they live in these sets—these ecological dramas which literally shape the next time slice. Then you have these huge disruptions, these mass extinctions, that throw everything out of kilter.

"Hey, this bone is hard. Does anyone have a brush?"

A new bone has just exposed itself to Bob while he works on the ilium, and I ask if he can tell what it is.

"This is a rib. There's a tradition among dinosaur paleontologists to be costeophobes. They dig through ribs. Throw them away. There's this myth that ribs are just a pain, and tell you nothing. That turns out to be incorrect. Because ribs tell you about, among other things, breathing, and how big your cardiac chamber is—all dinosaurs had huge hearts—and how your shoulder blade was hung on your ribs.

"I'll give you a quick anatomy lesson. See this mark here? In that groove was the intercostal muscle. Those are the muscles that make you breathe. So you can reconstruct the breathing movements of dinosaurs with great precision."

I ask him to explain the "steel ceiling" for Mesozoic predators—one of his latest theories.

"Well, mammals have never succeeded in producing a predator bigger than a polar bear. At a thousand pounds [454 kg], that's the limit for mammals. The cat limit's about six hundred pounds [272 kg]. The biggest saber-tooth, the biggest tiger, the biggest lion . . . about six hundred pounds. All through history. So there are built-in limits to each family. And for some reason, which is not at all clear to me, not only do mammals not get seven thousand pounds, they could never even, as predators, get a ton. That's peculiar.

"If you follow the success of dynasties of predators like this megalosaur, one of the earliest ones to reach seven thousand pounds [3,178 kg], he's much shorter in the leg than *T. rex*. Much longer in the torso. Lower to the ground. Certainly more maneuverable, but not nearly as fast. Head just as long, but not as wide. It has teeth as big as a *T. rex*, but the bite's not as strong. *T. rex*'s whole anatomy is concentrated to two functions. Running—it's got very long legs for its size—and biting. That's the two things it does.

"*Rex* is the ten-thousand-pound roadrunner from hell.

"Write that down," he says. "They'll like it."

"Turning? Not so good," he critiques. "Because the body's so stiff and birdlike. The whole history of these big predators—what I call 'Rivals of *Rex*,' the over-seven-thousand-pound predators—every time one family goes extinct and another one evolves, the new one is more birdlike."

"In what way is a *T. rex* like a bird?"

"Well, in what ways is it *not* like a bird? Okay, it didn't have wings, but both backbone and leg bones are hollow. The neck bones are hollow like a bird's. The skull bones are hollow like a bird's. The nerves exiting the brain are like a bird's. Marsh pointed that out in 1880. He didn't have a *T. rex*, but he had an ostrich dinosaur, which is very close. He said, 'Cripes! This looks like a bird.'

"These groups have a great number of air pockets built into the braincase bones. It's not just that it made them lighter. It's the air circulating between the jaw muscles and the brain, which among other things prevented the brain from overheating. The holes going to these air pockets in a *T. rex* are about this big." He forms his hands into the size of a large orange. "Like conduits. And in birds that air sac system in the head is connected with the whole rest of the body, which has several functions. It makes the skull lighter and stronger. And certainly prevents the brain from overheating."

"But the brain wasn't that big, was it?"

"That makes it even more dangerous," Bob says. "The smaller the brain is, the more quickly it can overheat. *T. rex* does not have a big brain for its size. But in the *T. rex* you have a big predator that's out in the middle of the day in a hot environment killing prey. It's going to generate a lot of heat with its jaw muscles. And the outside heat from the sun is potentially very dangerous to the brain. But that heat can't go directly to the brain because there's this separation layer circulating air which might even have been cooled by evaporation from the throat. Now if that's true, that's refrigeration. That's a refrigerated brain."

Bakker is known for finding the smallest known plant-eating dinosaur, the one he calls *Drinker* after the famed bone hunter Edward Drinker Cope. (See "What's in the Box?" p. 15.) I ask Bob how *Drinker* was found.

Bob points to Jim. "This is one of the two original Jims," Bakker says, reintroducing his assistant slaving away in the next trench like an overlooked hero. "You've heard 'The Saga of the Jims'?"

Jim himself, sounding mystified, repeats, "'The Saga of the Jims'?"

Bakker begins, "Well, we're told by the government that amateurs have no business looking for dinosaur specimens. So, despite the government's warning, they stumbled through a Wyoming blizzard up and down arroyos, nearly killing themselves. Suddenly this shaft of light came down from the heavens and pointed to this one spot that was not covered with snow. And it was the first specimen of *Drinker*."

Jim laughs.

"Well," Bob says, "I'm embellishing only slightly."

"This story gets better every time I hear it," Jim says with a snicker. "I figure the next time we're going to have a pillar of fire."

"Can you describe *Drinker*?"

"*Drinker* as an adult is the smallest plant-eating dinosaur known. It is about the size of the embryos people find in eggs. And the smallest baby we have found has a thigh bone about that big." He pinches his fingers together, leaving a gap only a fly could navigate. "Really tiny. Well, we've got the front of the mouth. It's very strange. Most plant-eaters, the front of the jaws are wide, square-cut like a cow's. Most plant-eaters, big ones, have wide, squared-off faces for chopping off large numbers of leaves all at once. *Drinker* is not like that. The jaws come to a point like a dog's, or like a primate's. So they must have been pulling in leaves from the side of the face. Which is unusual. There's virtually no beak in front. The individual teeth might be the most complicated dinosaur teeth known. Lots of little cusps. The smaller, younger ones could have been eating insects. When you are that small as

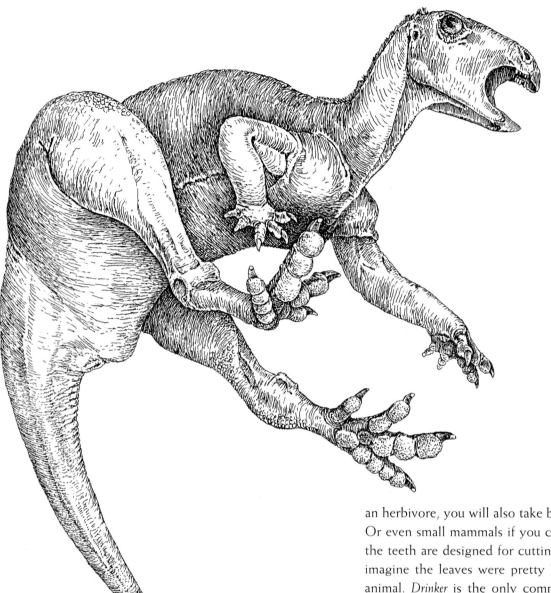

Drinker, a dinosaur from Como Bluff that Bob Bakker
named after famed dinosaur hunter Edward Drinker Cope.
Illustration by Bob Bakker

an herbivore, you will also take bugs. It's easy protein.
Or even small mammals if you can catch them. Well,
the teeth are designed for cutting leaves. But I would
imagine the leaves were pretty big compared to the
animal. *Drinker* is the only common dinosaur at that
age, and it's so small.

"It's all through these swamp beds. *Drinker* has big
feet. Huge feet. That's why we called it 'Little Big Foot.'
He has four big, spreading toes. I mean they're huge.
Stegosaurs have really, really short toes for their weight.
Drinker can walk across the mud without sinking. The
stegosaur, if you startle it, will sink. If you go to Southeast Asia, there are tapirs that live in moist forests and
swamps. They have spreading toes. So the feet should
correlate with the dinosaur's environment, and they
do. We find *Drinker* in the swamp beds right below
that sandstone cap called 'the Breakfast Bench.' It's this
huge swamp that covered thousands of hectares full
of fish."

" 'Breakfast Bench'? Is that the name of a formation?"

"The first place we camped, before we found fossils there, it was a convenient natural ledge for the Coleman stove. It was while adjusting the Coleman stove, sort of digging out some rocks so it would fit better, I found the first fossils. That's a true story with no embellishment."

I ask if every story should be verified for embellishment.

"If it didn't happen that way, it should have. Another interesting story that doesn't require embellishment: A reporter from *Discover* magazine was here from Montana. He was out to get the ambience that you writers do. 'Oh yeah! Can I get the ambience of all your work and your whole study to which you have dedicated your life and do it in an afternoon?' "

I didn't tell him, but true to form I was leaving that afternoon.

"So this guy works out of Montana and did a story on Elizabeth Claire Prophet, that religious cult up in Montana, just north of Yellowstone. These are mean people. They have automatic weapons. Anti-tank weapons. They have underground shelters. And they don't like the press and they put curses on reporters. They start chanting. They'll get their thousand, five thousand believers all chanting that you break your ankle. And two of the reporters who wrote negative pieces, strangely enough, had ankle injuries. And people were getting a little spooked. Anyway, so he's up here, and he says, since we weren't finding much, 'Well, we should chant.' So he invented a chant to find stuff. We were talking across the Breakfast Bench, and I was telling him that we'll go for two, three, four, five years without finding anything. So he invented the chant and we're all chanting, 'Teeth and jaws and vertebrae, appear before my eyes today.' We kept doing that in a rhythmic way.

"Anyway, so he picks up a toe bone. It's a *Drinker* toe bone. Just a corner of one. Not a whole toe. But it's kind of neat. Then this high school kid with us picks up a little fragment of a jaw. You hardly ever see those. Then the reporter picks up a vertebra. Well, that's kind of neat. That made us happy. And then Constance and I saw roughly three hundred *Drinker* bones in a very small area that were weathering out. So, I got a vert, I got a jaw, I got this, I got that, I got another jaw, I got

a foot, I got a pelvis. And this was a spot I had checked on and off for the last twenty years and had found virtually nothing."

"Now whenever you see Bob wandering around, it looks like he's mumbling," Jim said.

Bakker chants, " 'Teeth and jaws and . . . ' That's a true story with no embellishment."

Drinker, admittedly, is unusual for traditional taxonomic nomenclature. I wonder who officially names dinosaurs?

"Generally, the person who first finds a specimen has the honor," Bob tells me. "Just for record-keeping purposes, it's much easier to remember a name than a number. And usually it's some relative of the person doing the naming. Not always, though. We'll have a news release about this quarry pretty soon and we'll call this megalosaur 'Big Ed.' "

" 'Big Ed'?"

" 'Big Ed.' I like short names."

"After?"

Bob breaks into the theme song for "Mr. Ed," the popular early-sixties sitcom that starred a talking horse.

"I thought dinosaurs had to have Latin or Greek names?" I ask.

"Nope," he says proudly. "No rule books to name dinosaurs. It can be Urdu. It can be Hindu. It can be an anagram. It could be nonsense syllables—as long as they are not overtly insulting to some colleagues. Do you know that the very name 'dinosaur' is a claim jump?"

"No," I say, incredulous.

"Yeah, same group of critters that Sir Richard Owen called the 'dinosaurs' in 1842 had already been named six years before by Hermann von Meyer, a paleontologist in Germany. Meyer called them *Pachypoda*, which means 'strong foot.' Reasonable name. They had strong feet. Sir Richard Owen just ignored it."

I ask him how the name *Pachypoda* was overlooked.

" 'Dinosaur' is actually a better name. See, it's marketing. It's not whether you are first or brightest or best. It's marketing. People remember your slogan. Your product. Would you remember 'brontosaur' or *Opisthocoelicaudia*?

"Is that an awful name or what?" he says. "That's a Mongolian sauropod. 'Opisthocoelicaudia.' I mean, give me a break. At least the names I coin are short and memorable. 'Ed.' Or 'Drinker'—two syllables.

"I have a reverse Midas touch today," he says. "Everywhere I dig the bone disappears. Which is okay now, 'cause I'm trying to get around the ilium."

Bob's most memorable and durable contribution so far has been his theory for warm-blooded dinosaurs. Until the late 1960s, popular theory described dinosaurs as slow-moving, cold-blooded creatures ill-suited for survival in an evolving world destined to be taken over by better adapted warm-bloods—the mammals. As an undergraduate at Yale, Bakker began championing his heretical idea that dinosaurs *did* have active, warm-blooded, mammalianlike metabolism. His theory set off an ongoing debate that invigorated the field with scientific investigation, and even though many of his colleagues subscribe to his warm-blooded theory, recently some of his supporters have retreated from their positions, saying not all dinosaurs were warm-blooded. I ask Bob if he's changed his ideas.

"Warm-bloodedness is not one thing. Warm-bloodedness is what warm-blooded animals do. They evolve fast. The battle for hot-blooded dinosaurs has been won if you look at the general public, or *The New York Times*, or the *Golden Book of Dinosaurs*. In tenured academia that's something else again. A lot of those people haven't accepted the fact that the earth is not flat. There's a phenomenon called 'Meyerhoffia.' When the idea of continental drift broke in the United States, it had been accepted in Australia for fifty years. It broke in the United States in the 1960s. There was a guy called Meyerhoff who made it his business to go 'harumpf' at every argument about drifting continents. And he did this for twenty-five years. He published these long papers, supposedly refuting every argument about continental drift. And the net result was that at the end of twenty-five years, all the textbooks accepted continental drift."

Surprisingly, he admits he wasn't the originator of the warm-bloodedness theory.

"Slow Loris Russell, crazy Canadian, had been talking about it in the early sixties. And before that still, in the mid-1950s, this pair of Texans, Enlow and

Brown, published exquisite papers on the microstructure of dinosaur bone. And that was '57 and '58. I found those papers as reprints in the museum files. Now, I had taken courses in paleo a lot, and nobody had mentioned these papers. They're great papers. They proved that all dinosaurs were warm-blooded in '57! Nineteen fifty-seven, and no one talked about it."

I ask him if he thinks all dinosaurs were warm-blooded.

"It's like all birds are warm-blooded. If you cut a dinosaur bone it doesn't look like a big croc or a big turtle. It looks like a big bird or a big mammal. The fundamental flaw of the cold-blooded arguers is the *Brontosaurus* argument. It goes as follows: '*Brontosaurus* was so big, it didn't have to be warm-blooded. It didn't need a high metabolism.' And you say, 'Well, that sounds reasonable.' But you've just proven that elephants are cold-blooded. Because they live in a warm environment, and they're big. Whales must be cold-blooded. They are bigger still. But they're not.

"The people who want *Triceratops* or *Brontosaurus* to be cold-blooded because they are so big ignore the fact that if you extend the argument, you're going to have to say, 'And elephants are cold-blooded. So is the white rhino.'"

"So there were no lukewarm dinosaurs?"

"There might have been some that hibernated. Occasionally you find growth rings in teeth and bones. And you find growth rings in the teeth and bones of mammals and birds that hibernate, or go through a severe dry season, like lions. Cessation of metabolism can occur if your food supply is real low or if you are hibernating. If you section the tooth root of a black bear, there are very discrete growth rings for every winter they hibernate. That's how wildlife managers age a bear carcass. If you get somebody you think killed an underage bear, you pull one tooth out. Section it. Stain the section. You can count the growth rings. You can do that with lions, too, but with them it's not hibernation. It's that in the dry season they don't feed well. And they stop growing. You can do the same with hyenas."

Unlike bears, hyenas are pack animals—they hunt in groups—and I ask Bakker if he thinks predators like *T. rex* ran in packs.

"It would certainly make hunting *Triceratops*, which is a dangerous animal, a lot easier. Most of the prey animals were horned animals. I'm sure a lot of tyrannosaurs went home licking their wounds."

I cite a report by dinosaur tracker Giuseppe Leonardi, who says he found evidence of about fifty large predators traveling together.

"Those could be packs. Or breeding aggregations. Or migrations, like hawks. By the way, this is probably the best place in North America to watch predatory birds. During migration you can see ten thousand hawks here.

"Like the arrows of the Persians at the Battle of Thermopylae," Bob continues. "Remember what the Spartan general said when he was told that the Persian arrows will darken the sky? He said, 'Good, then we'll fight in the shade.' He spoke like a real Marine."

Spartans? Arrows? Ah, shields. Hmmm, plates—*Stegosaurus* plates. "What were *Stegosaurus* plates for?" I ask.

"It's partly for defense, but it's mostly species recognition. Different stegosaur species have different-shaped plates. Some are triangular. Some are triangular with a point on top, and the shape difference is very obvious. And that's standard in animals today. I mean, antelope horns are partly for defense. But also antelopes can tell each other apart by looking at the horns."

Jim asks if they were used as thermal adjusters.

"Anything that sticks out of your body can be used to dump heat," Bakker says. "A moose antler, when it's in velvet and there's blood flowing over it, they can use them to dump heat. But the main purpose of a moose antler is to intimidate other male moose."

"Same thing with the horns of a *Triceratops* then?" I ask.

"Well, not those horns. Those horns are deadly, and the horns over the eyes are really deadly. They would have been this long," he says, stretching his arms out like a fisherman telling a good story.

I ask him who he thinks was the fastest dinosaur.

"Depends whether you are talking about acceleration, distance running, broken-ground running. It's like a cheetah is extremely fast but poops out immediately and is only good on flat, level ground. Antelope can keep up cheetah speed for much longer than local prongbuck can. Lions accelerate extremely fast but

their top speed is only modest. Wolves have a very high sustained speed."

"Could a *Tyrannosaurus* sprint thirty miles per hour?"

"More. Once it got started. It's very long in the shin and ankle. Once I was describing a pterodactyl so I was interested in knee ligaments, and it had marks on the shin and the thigh for the knee ligaments.

"No one had diagrammed the knee ligaments in a bird very well, so I dissected fifteen chickens. Now I am the world's authority on knee ligaments in chickens! It's exactly like a *T. rex* knee. You can see where the ligaments attach. Kentucky-fried dinosaurs."

I ask Bob to describe other attributes of *T. rex*, arguably the world's most famous predator.

"Testicles as big as pumpkins."

Then he goes on. "*T. rex* is exceptionally advanced. It may be big, but it's exceptionally advanced. Braincase. Legs. Pelvis. Very birdlike in the way *Troodon* is very birdlike. All meat-eating dinosaurs had excellent sense of smell. On the inside wall of the snout bones. There are thin, curved ridges, look like a bony cannoli, you know the Italian pastry, but extremely thin. Those are called 'turbinals.' Dogs have them. Cats have them. And some birds do. And it's to increase the surface area available for the sensory cells that pick up the sense of smell. Well the snout of *T. rex* is gigantic. So the area available for the sense of smell is astronomical. It would have been like a whole pack of wolves put together.

"Hearing was pretty good, but not in the high frequencies. The spiral of the cochlea isn't as long, so that the area available for the sensory hairs is much smaller. Inferior to modern birds and mammals. It's the one sense system where dinosaurs are inferior to most birds. Sense of smell much better than an eagle, eyesight is probably as good. So we're talking about an eagle-eyed, wolf-snouted, ten-thousand-pound [4,540 kg] predator. *T. rex* is faster than anything else. And it's taller, so its search distance is much larger. It had stereoscopic vision. Both eyes focused on the same object. It's rare in meat-eating dinosaurs. Megalosaurs do not have it. And allosaurs don't, either. Nearly all dinosaurs have huge eye sockets. And the optic lobes in the brain are quite large compared to most mammals. Dinosaurs were color-sighted, very obviously."

I ask him how he knows this.

"The only direct descendants of dinosaurs are birds.

"And all birds are color-sighted," he says. "The only dinosaur uncle that's related to dinosaur ancestors are crocs and alligators. And they are color-sighted. Lizards are distant relatives of dinosaurs, and most lizards are color-sighted. So the whole family tree, all the branches that are around the dinosaur branch, are color-sighted. Color blindness is actually quite rare. On the other hand, most mammals are color-blind. They are nocturnal or crepuscular, living in dim light situations. The exceptions are the higher primates, which are active during the day. Few dinosaurs came out at night. It's like birds. Very few are nocturnal, maybe just one percent. Dinosaurs were color-sighted, and they had to be. At least in the breeding season, they would have had bright colors for attracting mates. And their camouflage would have to be color-coded. A dinosaur in a dense forest would have to be bright green. One in a dry flood plain would have to be reddish brown. One color they wouldn't be would be gray. Gray is wrong for camouflage. You don't see a lot of gray birds."

Bakker talks rapidly, free-associating, jumping from topic to topic with the spring of a prize-fighter.

"The reason people get excited about *T. rex* or megalosaurs is that they are rare. On average you get one big predator per twenty or thirty herbivores. Some dinosaur faunas, it's one in a hundred. You go to the dunes of Mongolia and it's something like three *Velociraptors* for every five hundred *Protoceratops*. You see, I believe you should study the common animals. They are the glue. Turtles and plants.

"You can learn a lot about what dinosaurs are and are not by getting to know a turtle. Turtles are the *Untermensch* of evolution.

"That's what the Germans used to call the Poles and Russians. The *Untermensch*. You know, the lower classes of humans. The ones you didn't have to worry

about. Well, turtles don't get any respect, unlike dinosaurs, but in fact turtles are survivors. There's never been a worldwide die-off of pond turtles. Ever."

"Why?"

"Turtles are slow-growth. Emphatically cold-blooded. They do very well in swamps and lakes where dinosaurs did pretty poorly. Turtles and frogs have that in common. Whatever they're doing ecologically, it's just the opposite of dinosaurs.

"If you want to be a survivor in evolution—not a success, but a survivor—be a turtle. Be a frog. Because they are the movers and shakers in your ecosystem. Rare animals don't make much of an impact.

"Actually, when the dinosaurs went extinct, statistically it was no big deal. They were one or two percent of the fauna. We think it's important because they are big."

A few miles away a storm begins to roll in over the Sundance Formation, sending lightning bolts bouncing off the sediments of the last Jurassic sea in America. The site begins to darken.

I'm put to work, the interview no longer a valid excuse to abstain from the task at hand.

"If you would start working down here," Bob instructs. "There's a bone here which you gotta hit with the glue. This crumbly stuff on the surface is very salvageable. It soaks up the thin glue quite nice. There are also a lot of teeth here. Must have had about fifteen. They are hard to get out perfect because they are crumbly. The nice thing about them is they'll tell you who was chewing up this stuff. These carnivores, like crocs, would shed their teeth so rapidly. Every time they bit something a tooth or two would fall out."

"Here's one!" I say.
"Oh mercy, yes! Yeah. Look around, you should find the rest of it."

"Here it is, right here!" I say excitedly.

"It's a meat-eating dinosaur tooth! The very tip is missing. Glue the two pieces together. Lay it somewhere, and when it's dry, we'll wrap it up and mark its approximate position. Did you do chanting? 'Teeth and bones and vertebrae, appear before my eyes today. . . .'"

A THEORY WITH A KICK

"It's part of the curse of collecting—the really important finds are made on the last day of a field season," John Ostrom of Yale University tells me. "I remember that instant just like it happened yesterday. I even remember the thoughts that went through my head. I realized this is something I had not seen before and all I saw was a couple of bone fragments on the surface."

It was 1964, the last day of his summer field season near Bridger, Montana, and Professor Ostrom was getting ready to pack his team of students back east for fall classes. He was hiking across a steep slope in the badlands with his field boss Grant Meyer to investigate some skull pieces discovered by one of Ostrom's students, Bob Bakker, when he saw some fragments on the ground.

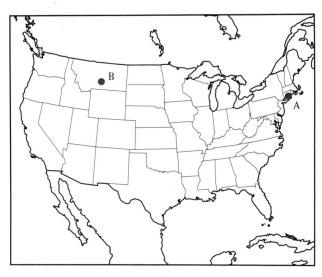

ABOVE: New Haven, Connecticut (A), and Bridger, Montana (B), *Deinonychus* dig site.
BELOW: The *Deinonychus* dig site dubbed "the shrine"
Photo by John Ostrom

Professor John Ostrom of Yale
University discovered *Deinonychus*, a
pack-hunting dinosaur that terrorized
victims during the Cretaceous with
sicklelike claws on its feet.

"From fifteen feet [5 m] away I knew that it was something different. I ran to them and collided with Grant because he was much closer than I was. But he hadn't seen it yet and BAM! I hit him in the shoulder and almost sent him tumbling down the slope. The bone was so different and so beautifully preserved, uncrushed, undistorted. . . . It was extraordinary. It was the shape of the claw, there was no other claw with that particular shape, that particular design. Without any hesitation, I said, 'We're gonna take it, right now.' We had no equipment, just my hunting knife and Grant's sandwich knife. We probed that slope for the next five or six hours until dark."

Ostrom and his crew had to wait until the next year before they got more of the predator out of the conelike hill which they later dubbed "the shrine." But even without further excavation, Ostrom had found the wicked adaptation that would give his creature its name.

Deinonychus, Greek for "terrible claw," had sickle-like claws on its foot. It would flick them out to rip open its prey like a kick-boxer with switchblades for toes.

"They were very active, agile, leaping animals," Ostrom explains. "They had sophisticated balance control with their tail, which was completely encased in bundles of ossified tendons, long rodlike structures used to make it more rigid, like the balancing pole of a tightrope walker or high-wire walker. Balance con-

trol was absolutely essential for their unique mode of attacking and killing. My interpretation of the site is that a flock—if that's the right word—or herd of *Deinonychi* were assaulting a victim—a *Tenontosaurus*, a herbivore—which was probably about half the size of one side of a tennis court. It wasn't a giant or a record setter by any means, but it was big—maybe ten times heavier than each *Deinonychus*."

In the fight that followed, Ostrom says, four or five of the *Deinonychus* gang fell victim to blows delivered by the *Tenontosaurus*, which was eventually killed by the remaining *Deinonychus*. Assessing the Cretaceous crime scene led Ostrom to develop an idea he had been toying with since graduate school—that dinosaurs were warm-blooded.

"The discovery and analysis of *Deinonychus* was a clincher because all of its anatomy indicated an animal that was very active, much more active than we had

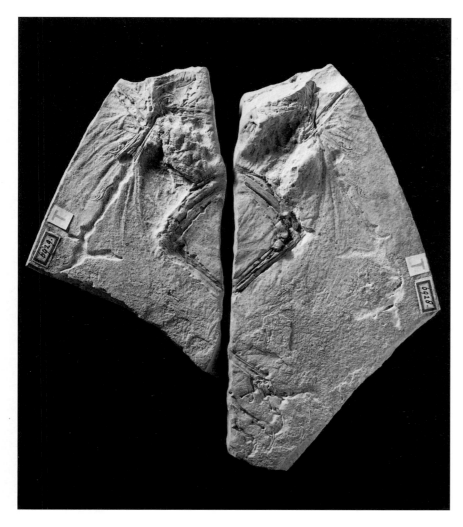

ogy was in graduate school. Dinosaurs have highly specialized dentitions in many of the ornithopods. That means that they were masticating their food for digestion to get nutrients into the bloodstream as rapidly as possible. That's the way in which mammals thrive—they have a high metabolism and need lots of food—not at all what you'd expect from a reptile.

"I had, of course, become fairly well acquainted with theropod anatomy as a result of finding *Deinonychus* and trying to figure out who he was and who he was related to. But then I had another fortunate moment. I was in the right place at the right time. I found a specimen of *Archaeopteryx* that had been misidentified for more than a hundred years. I was, at the time, also studying pterosaur flight and how it evolved. I went to see a type specimen of a pterosaur at the Teyler Museum in Haarlem, in the Netherlands. I was led to the gallery and shown the specimen under glass. They opened

been thinking of for dinosaurs. *Deinonychus* was an upright dinosaur, it walked on two legs. All animals that are sprawlers today are ectothermic—cold-blooded, and all upright animals are endothermic or warm-blooded—there are no exceptions and I felt that was a pretty powerful clue. If any of the dinosaurs were warm-blooded, *Deinonychus* was the prime example."

His warm-blooded theory, when published in 1969, sent tremors throughout the scientific community and a new dinosaur renaissance began. The warm-blooded dinosaur theory, however, is also largely credited to one of Ostrom's more celebrated Yale graduates, Bob Bakker.

"I think Bob and I reinforced each other," says Ostrom. "Bob is the one who made it newsworthy. I can't say when he first thought of warm-blooded dinosaurs, but I remember the first time it occurred to me that dinosaurs didn't have a simple reptilian physiol-

the case and left me alone to do my thing. When I picked up one specimen, identified as a pterosaur, I said to myself, 'It can't possibly be a pterosaur.' There are certain anatomical features that right away gave it away as not being a pterosaur. Then I noticed the label said the specimen was from Solnhofen, where all the specimens of *Archaeopteryx* are from, and the first thing I wanted to determine was, Is it an *Archaeopteryx* or not? Sure enough, I tipped it up to the light, low angle, and I could see the impressions of feathers. At that time, in 1970, only three other specimens were known. It was an extraordinary opportunity and it was a very frightening and embarrassing set of circumstances. The curator in charge provided me with a chair to sit down and take notes. After I had finally established to my own satisfaction what it was, I sat down and of course my pulse rate had gone up skyhigh—I was in a state of euphoria—I guess they call it

In 1877 O. C. Marsh turned down an offer to buy the
Berlin specimen of *Archaeopteryx* for $10,000.
It is now priceless.

---

cloud nine. I thought, I've got to compare this with
the Berlin specimen in East Berlin and with the London specimen and the Maxberg specimen. I've got to
show my colleagues that this is really a bona fide
Archaeopteryx.

"I asked myself, Do I tell the curator what I think
this is? Or just ask him if I can borrow it? That was
a major dilemma because I knew it wasn't likely that
any keeper would loan his specimen of *Archaeopteryx* to
somebody else, no matter who they were. I told him

what I thought it was and asked him if
I could borrow it. He got all flushed
and disappeared and I thought I had
blown it. And about five minutes later
he came back. He had taken the two
slabs—you can hold each one in a
hand—and he put them in a battered
old shoe box with a piece of string
around it. He handed me the shoe
box and said, 'Here'—I quote him
exactly—'You have made the Teyler
Museum famous.' "

In his haste at the Amsterdam airport to get to London and compare
his *Archaeopteryx* to the type specimen,
Professor Ostrom left his suitcase and
briefcase, including the shoe box, sitting on the curb in the care of a busy
skycap while he went to return his
rental car. "I was about a quarter of a
mile away, on a one-way street, and I
went, 'Oh my God! I left the shoe box
there!' I was in a state of panic, having
a heart attack a minute until I got
back to that counter. I'd only had it
for less than twenty-four hours. It was
still there.

"I stopped my enthusiastic pursuit
of pterosaurs and got into the bird/
dinosaur question. To call birds 'dinosaurs,' I think, obscures what we really
want to convey to the lay audience. We have a pretty
good understanding of what a dinosaur is with regard
to certain anatomical characteristics. A bird has those
characteristics plus some of its own which automatically make it a bird. Now if you're going to subsume
birds into dinosaurs, you're going to alienate every
ornithologist on earth. Why obscure them? Are you
going to go out and dinosaur-watch this morning? We
don't call ourselves reptiles, even though the reptiles
ultimately gave rise to some stock which gave rise to
the most primitive grade of mammals. Why is it necessary to confuse the question by saying they're dinosaurs? A bird's a bird. And it came from a dinosaur-type
ancestor. Specific ancestry is very much in question.
I'm not claiming *Archaeopteryx* is on the main line of

avian evolution, but I don't see any characteristic, any anatomical feature in *Archaeopteryx* that makes it impossible to have been an ancestor of later birds. I think the simplest explanation is that the two are related."

Ostrom later learned that if the price had been right, he might well have found himself the curator of the Berlin specimen of *Archaeopteryx*, arguably the most valuable fossil in the world.

One day in August 1983 Ostrom found a manila envelope on his desk from the librarian of the Yale Vertebrate Paleontology Library, just next door to his office. Inside was a folded-up tissue paper drawing. "I unfolded it very, very carefully, because it obviously was brittle. And there was a tracing of the Berlin specimen of *Archaeopteryx*—the one you think of first. There was a letter with it—the text was in German script, so it was hard to translate it. It was written in 1877 by some dealer in Nuremberg, Germany, writing to Professor Marsh at Yale—offering him the Berlin specimen for ten thousand bucks! My first reaction was, *That idiot! He didn't bite.* He had a reputation of being a penny-pincher. And then I thought, *Oh my God! That specimen is clearly one of the most prized and one of the most sought-after specimens in the world.* I was grateful that I didn't have the enormous responsibility of house-sitting that old bird."

DINOSAUR CHIROPRACTOR

Peter Galton was telling me how he had straightened the back of nearly every known dinosaur, but I was having trouble listening. As he spoke, there were several small aquarium tanks of human brains behind him whose owners were lying on rows of tables in the next room. At my request, we went to a more private room to talk.

Peter Galton is a paleontologist who now teaches anatomy to chiropractors in Bridgeport, Connecticut. In the 1970s his straightforward logic of anatomy compelled nearly every major museum in the world to begin the daunting task of realigning their bipedal dinosaurs into the now accepted mode of locomotion: back and tail parallel to the ground and head slightly raised in a stance reminiscent of a large chicken.

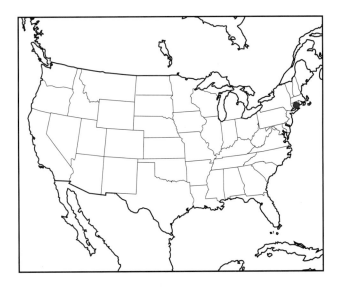

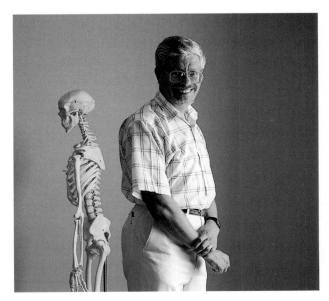

TOP: Bridgeport, Connecticut
ABOVE: Peter Galton is a paleontologist who teaches anatomy to chiropractors in Bridgeport, Connecticut.

"I started looking at the logic of upright dinosaurs and realized it didn't make sense," he says in one of those crisp British accents Americans love. *"Iguanodons* were mounted with their tails bent, but that was done by basically breaking the vertebrae so they would fit the mount. They have a tail that's probably as long as their body, so to walk upright they would have to break their tail and put a skate under it. Dinosaurs are basically horizontal."

ABOVE: Mesozoic bonehead with modern head-banger gear.
Helmet courtesy of the Dubuque High Rams
OPPOSITE BELOW: Boneheads depicted in courtship ritual
Illustration by Bob Bakker

Thinking of horizontal dinosaurs got him thinking about what the boneheads might have been up to.

One of the most unusual dinosaurs ever found is the scholarly-looking *Pachycephalosaurus*, or, to use the paleontological vernacular, the boneheads, which are only known from the remains of their large knobby heads that litter the Cretaceous sediments around the world. "Have you ever seen a skull of *Pachycephalosaurus?*" Galton asks. "Its closest nonliving relative is a cannonball. Their heads are eighteen inches [46 cm] long, eight inches [20 cm] thick, and it's solid bone." At first glance, the boneheads give the impression they were intelligent creatures, but Galton believes their specialized headgear evolved from physical confrontations stemming from a courtship ritual.

"If you've got a horizontally held vertebral column and you put your head down—then you've got an excellent battering ram, so I came up with the idea that boneheads were essentially modern equivalents of the mountain sheep that are found in Alberta. They would run toward each other and put their heads down at the last moment and ram. The winner got the girl. Bob [Bakker] did a fresh restoration that's probably the most copied drawing, with variance in every kiddy book on dinosaurs. You cannot pick up a book on dinosaurs without seeing a picture of two head-butting boneheads—which is funny because the rest of the skeleton of *Pachycephalosaurus* is unknown. It's only known from its skull."

"Any other paleontological claims to fame?" I ask.

"I put the cheeks back on dinosaurs," he says proudly. "In the 1940s dinosaur cheeks went out of style when Brown and Schlaikjer, in their *Protoceratops* monogram, pointed out that only mammals have a crucial muscle for cheeks. Just about everybody agreed with the arguments. But the mechanics of the ornithischian jaws are such that the teeth face outward and upward, and unless you have different types of gravity back then, food would roll outwards and out of the mouth if they didn't have cheeks. It never made sense to me that these dinosaurs would go to all this trouble to cut up their food to merely just chuck it out. You'd see these ornithischians going along with bits of food ejecting from both sides of their mouths. From a functional point of view, you really do have to have cheeks. So in the 1970s cheeks became fashionable again."

Wisconsin Johnson Stops a Herd of Galloping Dinosaurs

Sixty-five million years ago the Hell Creek Formation in Montana was home to some of the last large dinosaur herds on the planet. Those great horned dinosaurs, the ceratopsians, are imagined to have covered the landscape like buffalo did in the old West. *Torosaurus*, among the largest of the horned dinosaurs, was an animal known only from its enormous skull before a crew from the Milwaukee Public Museum began collecting there.

Rolf Johnson, the museum's curator of paleontology, recollects, "I went out in the field with a crew in 1981. We actively use volunteers in these digs and we let them know that, gosh, everybody thinks they're going to find a big skeleton, but sometimes you'll walk for days and days and you'll be lucky if you find a small scrap of dinosaur bone. On the first morning of the expedition, literally in the first couple of hours, two of our volunteers, Bob and Gail Chambers, were walking alongside a hill and decided to go to the top

BELOW: *Torosaurus* dig site, thirty miles (48 k) southeast of Ft. Peck Reservoir, Montana (A), and Milwaukee, Wisconsin (B).
OPPOSITE: High-fidelity casts, universal joints, and elastic bands duplicating muscle attachments led Rolf Johnson to the regrettable conclusion that ceratopsians could not gallop.

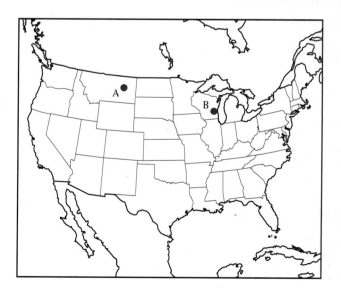

ABOVE LEFT: An upright stance for a ceratopsian's forelimbs is a position Rolf Johnson declares impossible, given his interpretation of its anatomy.
ABOVE RIGHT: A sprawling stance for the front forelimbs, a position he believes is more likely, would have made it impossible for a ceratopsian to gallop.
Illustration by Pat Redman

of it. Halfway up the hill they ran into a rattlesnake, so they had to find another path to the top and stumbled across this skeleton. We immediately field-identified it as *Triceratops*, which are a dime a dozen—very, very common.

"When we got it back to the laboratory and began preparing it, we quickly realized something was wrong. Some of the bones we thought were parts of the hip were actually the frill of a *Torosaurus*. We knew right away, *My God, we've got something special here*. We actually have a body, the first body of a *Torosaurus* ever discovered.

"We had always assumed the bodies of *Torosaurus* and *Triceratops* were more or less the same—kind of a conservative ceratopsian body plan. In fact they are

very similar, and now, by extension, we can also talk about the entire suborder of all the ceratopsians. And that's exciting.

"It's a real compelling image to have a herd of ceratopsian dinosaurs, you know, three to seven tons and maybe thirty feet [9 m] long, actually galloping, thundering across the Cretaceous plains. Some of the early arguments that Bakker and others had made for galloping dinosaurs was driven not so much by having good hard evidence, but because it was consistent with this theory that all dinosaurs were endothermic, high-metabolic animals. This is a real appealing idea and I had bought into it. I thought, *Yup, when we mount our* Torosaurus *here, those front legs are going to be right underneath the body.* Well, I tried to put them directly underneath the body, and it wouldn't work. Something was wrong.

"I had been very fortunate, because I was working with John Ostrom at Yale University, since two of the *Torosaurus* skulls are located there. I mentioned to him that something was going on with the forelimbs. He had never really bought into this placement of the forelimbs and he said, 'How would you like to shed light on this big debate dealing with the biomechanics of ceratopsian forelimbs?' And I just jumped at it.

"The real bones of course are very, very fragile, very brittle, and very heavy. So we made lightweight, high-fidelity casts of all the limb elements. And then we came up with an idea to mount this limb in kind of a large gantry and articulate these bones, put them together with universal joints substituting for the actual joints of the animal. We could then, in effect, take this front leg and put it in any position that we wanted to and see how the bones fit and if they worked in various positions. The bones were beautifully preserved, so we were able to determine where all of the major muscles were attached. I took large strips of elastic to duplicate the major muscle groups, to see if they would contract in a way that was reasonable to make the limb move, make it walk.

"There is no question that it is impossible, biomechanically impossible, to get these front legs in a parasagittal plane, an upright plane. Consequently, if galloping demands or almost demands having limb excursion in that upright plane, then *Torosaurus* could not gallop. It's a pity because the galloping ceratopsian mounts would have been pretty sexy."

A Microscopic View of Giant Bones

In 1969 Armand de Ricqlès, a professor of comparative anatomy at the University of Paris VII, quite separately from John Ostrom and Bob Bakker published a warm-blooded theory of dinosaurs based on his microscopic research of bones. We visited Armand at his comparative anatomy lab at the University of Paris.

"Particularly as a tissue, bone itself hasn't changed since the early beginnings of vertebrates," he tells us. "You can think of bone as something built of basic bricks, which are the bone cells. With bricks you can build castles, or little houses, or whatever you want, and this is what changes during evolution.

"When you look under the microscope, dinosaur bone tissue has a particular pattern, with a lot of blood vessels within it, and the bone among many dinosaurs just looks like the bone of fast-growing large mammals

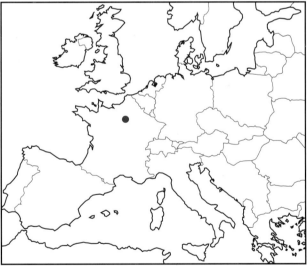

ABOVE: Armand de Ricqlès of the University of Paris came up with a warm-blooded dinosaur theory at the same time as Ostrom and Bakker by looking purely at microscopic thin-sections of bones and realizing dinosaur bone was just as vascularized as mammal bone.
BELOW: Paris

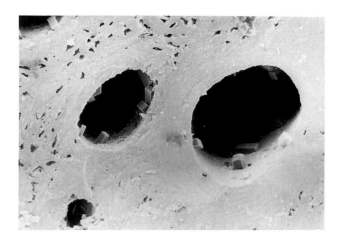

ABOVE: Blocks of quartz crystals form on vascular canals where dinosaur blood once pumped life to this duckbill dinosaur found on the North Slope of Alaska. For some reason there is very little perminerization of this bone, allowing researchers to extract some fairly intact DNA code. RIGHT: A duckbill dinosaur called *Hypacrosaurus* grew from embryo to full size in about four years. Ricqlès believes it was the dinosaur's overdeveloped hormonal system that allowed it to grow so large.

today, for instance camels or deer or cows or the big terrestrial birds like ostriches, which grow awfully fast.

"The idea that dinosaurs might have been warm-blooded, whatever it means precisely, is not new. It's a very old idea, which crept time and time again into the older literature. It began probably in Huxley's time, in the late 1800s. Huxley was the first who put dinosaurs roughly as intermediate between reptiles and birds. Something that was probably new was to bring histology into the picture.

"Histology is part of the biological sciences which study tissues. In our body there are more than two hundred fifty different kinds of cells. And all those cells self-produce and integrate into organs. Bones are specialized connective tissues, very akin to the one which forms the dermis of our skin.

"There is nothing special about the way dinosaurs were growing. The processes they used are quite straightforward—it's exactly the same way in deer or cows. To understand how they got so big, you have to think of the hormones, the hormonal system of the animal, controlled by the pituitary gland. It's possible to make molds of the neural cavity with the brain to have some ideas about the size of the pituitary, which is relatively large in many dinosaurs. Most dinosaurs probably were able to grow most of their lives. The large ones probably retained a reptilian growth pattern in the sense that they were able to continue to grow their whole lives. The older, the bigger. Of course there is also something I can comment upon for a few seconds; the famous problem of growth rings. First looking at dinosaur bone, I didn't see growth rings in my material. But in the older literature I had a few hints to the fact that some growth rings had been discovered. So I looked a little bit more in detail at the growth series of sauropod dinosaurs.

"You think of the growth rings in tree trunks. It's the same phenomenon with some dinosaur bones. There are yearly cycles in bone deposition, and this works as a kind of biological recorder. During the good season, in a tree, you have fast growth, and during the winter you have cessation, stoppage of growth, and this makes a kind of circumferential line. And the next spring growth is resumed and so on. You have one line a year. You have growth rings in some hard tissues of mammals, not just hibernating mammals. Even among the so-called cold-blooded animals, even if they are living under equable conditions, like the tropical regions where you have almost no seasons, you have external cues. It may be a tiny change in the duration of days, or a tiny change in the aridity; more or less water can be enough to trigger a cycle in bone deposition. In many animals it's possible to reliably use such rings to assess the age of an individual animal in years."

Since 1978 Armand has been studying dinosaur growth rates with Jack Horner, the Montana paleontologist who discovered a *Maiasaura* population, from embryos to adults.

"With Jack we found that the adult *Maiasaura* took less than ten years to become full adults, probably about seven or eight years. Ultimately it will be possible perhaps to answer the question of how long they lived, but for now we have little evidence. But it's not necessary to assume they had a very long life span in order for the dinosaur to reach that size. You could find papers from the 1930s assuming it took five hundred years for such dinosaurs to reach adulthood. I have nothing against the idea that they were very long-lived, but we don't have to assume long life to understand how big they were, that is the point.

"On the sauropod, I have a growth theory; I think, as a minimalistic assessment, it would be fair to think that perhaps it reached two-thirds of adult size in about thirty-five to forty years at most. So even for a sauropod, it isn't necessary to assume a centuries-old animal before they can reproduce. They would be senile before that.

"I am convinced, down deep into myself, that the dinosaurs were physiologically different. They were not what we see among the living world—mammals and birds or what we see among crocs or lizards or turtles—it was something in-between, I think. But I just cannot see a living dinosaur with the physiology of a large cold-blooded land turtle. To me it doesn't hold."

OUT ON A LIMB?
JACQUES GAUTHIER FINDS BIRDS IN THE DINOSAUR FAMILY TREE

"Dinosaurs didn't go extinct," Jacques Gauthier says. "In fact, they're doing quite well in the form of birds. *Triceratops* doesn't look like a tweety bird, but they are still related."

At first, Jacques Gauthier of the California Academy of Sciences in San Francisco wasn't really interested in dinosaurs. He was studying the characteristics of several lizard groups when he got sidetracked by what he calls "the curse of cladistics."

Cladistics is the method scientists use to determine family trees. Gauthier tells me, "Cladistics requires that you know not just one but at least two outgroups [nearest relatives] of the group you're interested in investigating. You have to know its next nearest relative and the nearest relative after that to determine unambiguously the direction of change. This leads to quite a structured tree and ever-widening areas of investigation [the curse], and that got me off into the dinosaur thing."

He goes on, "Our system of classifying animals was essentially given to us by Carolus Linnaeus, an eighteenth-century Swedish botanist who used a system of folk classification that extends back to Aristotle. Linnaeus defined animals and plants by the overall similar properties they possess. The problem with that system is that those properties he chose are inherently nonevolutionary. Take, for example, a plant-eater—it's called a herbivore. It begins to eat meat. You don't call it a herbivore anymore. You call it a carnivore. By classifying it as a herbivore, you have effectively for-

Ornithomimidae
Caenagnathidae
Deinonychus
Archaeopteryx
Hesperornithiformes
AVES

BIRDS: A FAMILY TREE

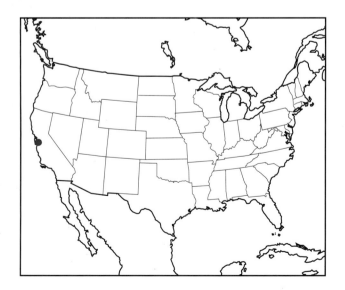

ABOVE: Jacques Gauthier, a pioneering dinosaur cladist, says that dinosaurs are not extinct but kicking ass in the form of birds.
LEFT: California Academy of Sciences in San Francisco, California.

bidden it to evolve with respect to what it eats. So classes cannot evolve, certainly not with respect to its defining characteristic. Evolution requires that things change indefinitely.

"There is any number of ways in which you could categorize objects in the natural world," Jacques explains. "You could put them into the group of things that I like or the group of things that I don't like. The group of things that I eat or the group of things that eat me. Things I find in the Southern Hemisphere or in the Northern Hemisphere. Above the ground or below the ground. There's absolutely no limit, but when you want to do a study of phylogeny—the genealogy of life, who begat whom—then you have to look for particular kinds of evidence to reconstruct that history separate from irrelevant data. The phase of the moon is data, right? You can look out there and see: Is it quarter? Is it half? Is it full? Right? But you wouldn't consider that data to be germane to the question, Did this person commit that murder? Maybe in California you would—but under normal circumstances, you wouldn't. Even though it's data, it's just not relevant to the question you are asking.

"So given that the question I'm asking is, How are these animals related to one another? there are only certain kinds of evidence that are appropriate. Many years after Darwin, not until the 1950s, in fact, did Willi Hennig, a German insect worker, consider the problem. He had a really great insight: he realized that some animals may not look very similar but must be related genealogically. Whales don't look very much like bats, but nonetheless, they are both mammals. Their overall similarity wasn't essential in determining a relationship; specific traits were. Both bats and whales have hair, are warm-blooded, and the females produce milk.

"You'll note that you have an opposable thumb and you also have hair," Gauthier explains. "But not every animal with hair has an opposable thumb. Right? Dogs have hair, but they don't have an opposable thumb. However, every animal with hair has a backbone. But not everything with a backbone has hair. A fish has a backbone, but it doesn't have hair. And everything with an opposable thumb has hair, and everything with hair has a backbone. So you can see that the attributes of any organism are interested within one another. That gives you a hierarchy. First animals developed backbones, then came animals with hair, and then came animals with an opposable thumb. That's the way generations of life evolved. The most distant generation contains the next generation, and that one contains the next generation, each with its own inherited characteristics.

"Every organism has characteristics which diagnose its relationships. All of those characteristics appeared at some point and enable you to reconstruct the history of life. Our problem is to find out when they appeared. Because only at that point do they diagnose a group, which enables us to recognize an organism as a unit of ancestry—an individual with a beginning, a middle, and an end that is spatiotemporally located—bounded in space and time. We are looking for individual units of descent that relate to a genealogical whole. Then you can use this method to reconstruct the history of life."

When Darwin wrote *On the Origin of Species* he knew he would have problems with what he called "transitional forms," or more popularly, "missing links," fossil animals that showed affinities to both their ancestors and their future relatives. In his 1859 treatise Darwin extensively used the fossil record, although he noted it may not be a reliable witness to evolutionary theory: "The crust of the earth is a vast museum," he wrote, "but the natural collections have been imperfectly made . . . "—explaining, he hoped, the lack of physical evidence for his controversial theory. Still, the poverty of missing links was seen as a weak link in his theory.

Upon publication of *On the Origin of Species*, a predictable cry was shouted from the pulpits of Victorian England denouncing Darwin's blasphemous theories. "Explain birds?" challenged the queen's outraged bishops. Birds, almost as much as man, were thought to be a divinely inspired creation. But birds, unlike man and his embarrassing resemblance to his jungle cousin the apes, could not be easily associated with anything else in the animal kingdom. They sort of just appear.

Darwin, never a public debater, recoiled from the controversy. Then one year later, in 1860, came a mysterious omen.

A single feather miraculously appeared in the Solnhofen Limestone Quarry in Bavaria. The quarries, once a large Jurassic backwater lagoon, line the Altmühl River and have been worked since Roman Empire times, when the fine-grained stone was used for tile in the Roman baths. Two thousand years later, in Darwin's time, the fine-grained stone was also discovered to be useful for lithographic printing plates, and that's when quarry workers discovered the feather. The following year, in 1861, the crown jewel of all fossils rose out of the same quarry system. It was one of Darwin's missing links.

Archaeopteryx is the first known dinosaur to possess
powered flight and is considered by most paleontologists
to be the first bird.
Illustration by Pat Redman

The Berlin specimen of *Archaeopteryx* is one of the most famous fossils in the world. Seemingly half dinosaur and half bird, it has been called a fossil caught in the act of evolution.

The discovery of a fossil feather in Solnhofen Limestone Quarry in 1860 was a prelude to *Archaeopteryx*, found one year later in the same quarry system.

Archaeopteryx, which means "ancient bird," was the transitional form Darwin and his supporters were looking for. Seemingly half dinosaur and half bird, it has been called a fossil caught in the act of evolution.

"Just two years after the publication of *On the Origin of Species*, wham, bam, thank you ma'am," Gauthier says, "out comes *Archaeopteryx*. It's clearly got the feathers of its descendants. But it still has a long bony tail and teeth of its ancestors. It was a triumph of predictive science."

Sir Richard Owen, Her Majesty's knighted paleontologist who had described *Archaeopteryx* in 1863 and later bought it for the British Museum, declared the creature to already be a bird and therefore not in transition.

Darwin, however, felt cautiously vindicated by this striking piece of evidence and left his friend naturalist Thomas Henry Huxley to champion evolutionary theory. Regarded as "Darwin's bulldog," Huxley unleashed himself for a zealous crusade against Victorian religious thought. At stake for the church was man's own unshakable place in the universe. Man must be regarded as God's finished masterpiece, not as a work in progress. An unbridgeable gap had to exist between man and apes. Defending man's hallowed place above the animal kingdom was a bevy of bish-ops advised by Her Majesty's knighted paleontologist and the man who in 1842 coined the very name "Dinosaurs," Sir Richard Owen. Owen's list of man's sacred and unique characters (large brain, upright stance, and so on) were actually, as Huxley pointed out, all conspicuously displayed by our own evolutionary kin, the apes.

Gauthier continues, "Now Huxley, in 1867, had just published his big monograph of the evolution of birds. At the time, Huxley had only a few dinosaurs available to him and only partial remains at that. Nonetheless, as he put it, he took the leisure hours of the winter of 1868 and '69 and had a look at the problem and discovered that the intermediates between crocodiles and birds were the dinosaurs. He did a perfect cladistic analysis but nobody accepted it.

"It turns out that the chickens called Dorking fowl, raised in Huxley's part of England, also called Dorking, don't have the leg bones fused up like a normal bird does. And Huxley noted that it made them look exactly like a theropod dinosaur. This is a guy who probably played with his food, just like I do. At

Thanksgiving dinners I bore my family cross-eyed—the big roasted theropod dinosaur lying on the table there. I give them an anatomy lesson with all the details, and they just want to eat it. I can hear his wife telling him, 'For God's sake, just eat it—quit playing with it.' And him diddling up, chewing off parts, so he could see the bones better. In one of his papers Huxley said that if you took the leg of a half-hatched chick and blew it up to the size of a dinosaur, you wouldn't be able to tell the two apart. That was essentially his argument for birds coming from dinosaurs.

"When I got involved, there was just a ton of evidence to support that birds are dinosaurs. I wondered what the hell have we been fighting about, because this has been a long-term battle for over a century. The evidence points clearly to one conclusion: that birds are a kind of dinosaur.

"Just as man's past is deeply embedded within primates, likewise birds are embedded in dinosaurs," Gauthier says. "One line of dinosaurs took to the air and became birds. Basically I changed the taxonomy to reflect evolutionary relationships rather than these outdated class concepts. I actually put Aves [modern birds] in Reptilia because birds are a kind of reptile. Before it was an 'expert' who said that his group was related to another group because he thought they were similar. Period. Now, with cladistics, you have to say, 'I think birds are related to dinosaurs because . . .' and you have to be very specific about your defining characters.

"You make what's called a taxon character matrix, where you list the taxa, like birds and dromaeosaurs and allosaurs, and then you note, Do they have four fingers on the hand? or, Do they have three fingers? and go through each individual and score each characteristic. Then you can run your list through a computer program that crunches it all up and will show the internested relationships of these attributes and print a family tree out for you."

When Gauthier came up with 132 similar characteristics of birds that are shared also by dinosaurs and their extinct allies, the computer told him something that Huxley suspected over a hundred years ago: Birds unmistakably are dinosaurs.

"The battles of cladistics have been won," Gauthier says with conviction. "The war is over.

"By cladistic analysis, you can only say that birds are a kind of dinosaur. Dinosaurs run the entire range, from those bombers like *Ultrasaurus* and *Supersaurus* down to those little Cuban hummingbirds that weigh a couple of grams. There are about nine thousand different kinds of birds alive today and only about four thousand mammals, so there are twice as many dinosaurs around as mammals. That's always the way it's been. Go back two hundred million years and there were always more dinosaurs around—at least in terms of species diversity—so dinosaurs are still kicking ass. I don't know why they call this the Age of Mammals when it's still the Age of Dinosaurs."

LOST EPOCHS FOUND IN ASIA

CHINA: LAND OF THE TERRIBLE DRAGON

The first source of good Asian fossils came from Chinese pharmacies, where great numbers were being ground up and ingested for medicinal purposes. Max Schlosser, a doctor from Munich, reported this phenomenon in 1903 after another doctor named Harberer had amassed a large collection of Chinese fossils, all rescued from Chinese medicine shops.

Dong Zhiming, China's leading dinosaur paleontologist, took us to visit some pharmacies around the city of Zigong in the Sichuan Province where, like all over the country, fossils are still being pulverized into medicine.

"The calcium in the bone is thought to be useful for kidney ailments," Dong tells us after interviewing a doctor who had set up a desk on the sidewalk outside a medicine shop and was diagnosing ailments and writing prescriptions for a long line of patrons waiting to get in. We walk down the street to another medicine shop's doctor for a second opinion. "Headaches, too," Dong translates with a straight face. "He recommends the powder form and dissolving it in alcohol."

Dong, a student of the great Chinese paleontologist C. C. Young, tells us the Chinese word for dinosaur is *konglong*, which means "terrible dragon." The

ABOVE: Chinese apothecaries, like this one in Beijing, still sell ground-up dinosaur bone for pharmaceutical purposes. "Stone dragon bones" are believed to have the power to cure a variety of ailments.
OPPOSITE: At the Zigong Dinosaur Museum in the Sichuan Province, Chinese paleontologist Dong Zhiming studies the neck of a twenty-meter-long (65.67 ft) *Omeisaurus* from a bosun's chair.

Chinese, he says, have prescribed "stone dragon bones," or *lung ku*, to cure a variety of ailments since the twelfth century. He estimates that at least one hundred tons of "stone dragon bones" are ground up and used for pharmaceutical purposes throughout Asia every year. Most of the medicinal bones, he says, are from fossilized mammals of the Pleistocene, about five million years old. However, in 1976, while passing through the nearby town of Weiyuan City, also in the Sichuan Province, he tells us, he saw more than ten tons of dinosaur bones stacked up in a medicine shop, all from the nearby Jurassic Red Beds.

The whole middle of the Jurassic was pretty much a black hole in dinosaur research until the mid-1970s, when a road crew cutting a swatch through a hill for a new highway outside the city of Zigong discovered a virtual cemetery of Jurassic dinosaurs. The excavation, called the Dashanpu Quarry, eventually spread to over twenty-eight hundred square meters (111 sq rods). Dong, who led the excavations there from 1979 to 1981, says the quarry could extend to over twenty thousand square meters (791 sq rods) if a few apartment buildings that are in the way are torn down.

The Zigong Dinosaur Museum, erected directly over the quarry, opened in the spring of 1987 and is one

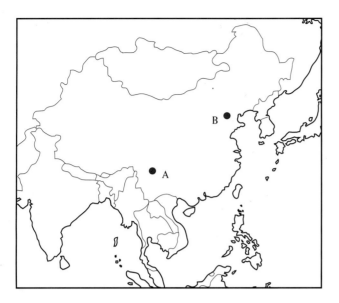

ABOVE: Zigong Dinosaur Museum and Dinosaur Quarry in the Sichuan Province (A), and Beijing (B).
BELOW: Josué-Heilmann Hoffet, circa 1930
Illustration by Pat Redman

of the most spectacular dinosaur museums in the world. Over eight thousand bones, representing at least one hundred individuals from eight different species, have been taken from this quarry since its discovery. Spectators can view excavations from several pedestrian bridges that ring and bisect the quarry. Many of the bones have been left *in situ*, much as in Dinosaur National Monument in Utah, and about a dozen mounted dinosaurs are on display in an attached museum—enough dinosaurs, Dong calculates, to keep Asia's pharmacies stocked with headache remedy for about a year.

THE SACRIFICE

Philippe Taquet, a professor and director of the Laboratory of Paleontology of the National Museum of Natural History in Paris, told us the story about his expedition into the little-known dinosaur territory of Indochina.

"Fifty-five years ago, a French geologist called Josué-Heilmann Hoffet went to Indochina to conduct a geological survey and found dinosaur bones. Near the border between Vietnam and Laos there is a very small village in a dry teak forest. The people of the village are Qatang and they are animists. Hoffet arrived in this small village, and when he began observing the rocks, the people say, 'Oh, you must be looking for the stone bones—of the sacred buffalo.' They told Hoffet that when the sacred buffalo are young, they carry the sun in the sky each day, and when they get old, they die in this place, not far from the village. This is their legend. They took him to see the big vertebrae of the sacred buffalo. They were dinosaur bones. Hoffet was fascinated and wanted to collect the bones, but they told him, 'You cannot touch these vertebrae because they are sacred. If perhaps you do a sacrifice, maybe you can collect some of them.' So he paid them for a buffalo, a young buffalo, and he was allowed to collect some of the fossil bones. He described these bones in 1936 in Hanoi in a small paper, 'Description of New Titanosaurians in Bas-Laos.' After Hoffet described the new dinosaurs, he was killed in 1943 by Japanese soldiers invading Vietnam during the Second World War.

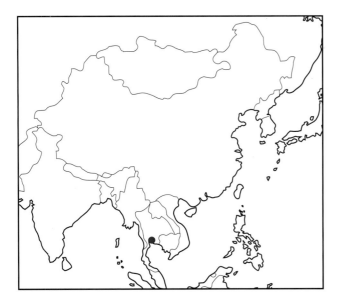

Site of "sacred buffalo bones" excavated in Laos by Hoffet and later Philippe Taquet, director of the Paleontology Institute of the National Museum of Natural History in Paris.

"I received a paper some years ago," Taquet says, pulling a yellowed document from his files. "It was a map Hoffet made. Here is Savannakhet, Mekong River; the Ho Chi Minh track was here. And he says, 'There are bones all around here, around this tiny village.' I had this paper many years, so when I finished my job as director [of the Paris National Natural History Museum], I said, 'Now I go.'

"*Now* it's possible to go back to this country, but it was impossible to go there for many, many years because of the war and the Russians there. Two years ago I went to Laos and I took this map. Laotians are very sympathetic and kind to the French, because they speak French, so they helped me. We went into the forest and I came to this small village. I explained to the people what was the reason of my trip, and they bring to me a very nice old man, seventy-five years old, called Mr. Kommapa. And Mr. Kommapa said, 'I

was working with Hoffet. I was carrying the boxes with the fossils. I know where the sacred buffalo vertebrae are, but you must pay a sacrifice.' Same story as Hoffet received.

"So I bought two pigs, and we killed the pigs to make a sacrifice and all the villagers ate the pigs. It was the first time I ever had to sacrifice a mammal to get a reptile.

"I found new localities with nice vertebrae. But I decided to come back the next year to collect the bones. I arrived back [in] Paris in December of 1990, and in January of '91 I gave a lecture in Grenoble on dinosaurs. I put in some slides of my last trip, to Laos to explain to the people the story of Mr. Hoffet. At the end of my lecture, a lady came to me and said, 'I am very impressed by your talk of Mr. Hoffet; I am his daughter.'

"She was a teacher in Grenoble. I found the two daughters and Madame Hoffet, who is eighty years old. And they bring to me an old black and white picture of Hoffet in the forest in 1935."

They had returned to France from Hanoi after Hoffet's death.

"I went back last year, but this time I had to buy a buffalo for sacrifice. It was more expensive, but we collected a nice articulated sauropod, so all is going well. And now the Laotian people want to make a museum in Savannakhet to show these dinosaurs."

I call Taquet, who has just arrived back from his yearly pilgrimage to Laos. "I had a fantastic time," he says. "Hoffet's son, Jean Fred Hoffet, came to visit us in the field. He was digging up dinosaurs side by side with Mr. Kommapa, who took his father into the forest sixty years ago. The dinosaurs we found are probably all new species."

Then Taquet adds, sounding relieved, "Now it is not necessary to make a new sacrifice for digging dinosaurs."

The Dinosaur Trackers

Besides bones there is another valuable and relatively unexplored tool for understanding dinosaurs' behavior—dinosaur footprints. While dinosaur tracks were actually discovered before dinosaur bones, they weren't recognized for what they were until much later.

Scientists in the first part of the 1800s mistakenly identified the tracks they found embedded in mud-turned-rock "stony bird tracks." Ever since the "bird tracks" were discovered to be "dinosaur tracks," the field has suffered from being called an inexact science. A hundred and fifty years after the word "dinosaur" was coined, scientific thinking has come full circle and most paleontologists now regard birds as dinosaurs.

There is now a small, enthusiastic fraternity of scientists quickly making the field into a highly regarded science and discovering remarkable clues to dinosaur behavior. Dinosaur tracking, for example, is the only way to determine an animal's speed. You can stare at an animal's bones until eternity and still not be able to calculate its speed. However, one tracker has formulated an equation, and tested its validity on living animals, to determine the cruising speed of dinosaurs.

The potential information from trackways is staggering. A bone hunter can get bogged down with the daunting task of preparing and describing the anatomy of one dinosaur for a good part of a lifetime; a dinosaur tracker can go out and in a few long weekends amass a whole regional dinosaur census based entirely on dinosaur tracks, because there are literally billions of them out there.

A group of juvenile brontosaurs
trek along the shore of a Jurassic lake
in southeastern Colorado.
Illustration by Douglas Henderson

IN THE FOOTSTEPS OF GIANTS

CRUISIN' THE DINOSAUR FREEWAY

Martin Lockley, son of a well-known Welsh naturalist, is a professor of geology at the University of Colorado at Denver who has been at the forefront of dinosaur tracking in this country for the last ten years. His mission in life is to elevate the overlooked art of dinosaur tracking to a respectable science. Lockley heads the Denver Dinosaur Trackers, a group that has been describing hundreds of track sites from Uzbekistan to the American West.

"It's ridiculous how many track sites are around Colorado," Lockley tells me as we drive south along Route 93 out of Boulder. "There's some right up here." He pulls off the side of the road and climbs over a fence and shows me the footprints left behind by some hadrosaurs that strolled across a mud flat one Cretaceous day.

As cars whiz by, he explains, "When we first started the Denver Dinosaur Trackers, we knew of a couple of dozen very sketchily documented track sites over the Colorado Plateau down into Utah and into the Four Corners area. And now after ten years of collecting data, actively looking for sites, we have some-

ABOVE: Martin Lockley with the cast of a thirty-four-inch (86 cm) *T. rex* footprint discovered by geologist Charles Pillmore of the U.S. Geological Survey in the Raton Basin of northern New Mexico. Lockley calculated that the dinosaur was moving at least 7 mph when it left its print some 65 million years ago.
OPPOSITE: Late Jurassic Santa Fe Trail: Lockley investigates water-filled brontosaur trackways illuminated by a sunrise over the Purgatoire River in southeastern Colorado. The parallel tracks, embedded along the shoreline of an ancient lake of the Morrison Formation, are the world's longest and are convincing evidence that sauropods were social animals.

where in the region of three hundred sites in our files."

"How many footprints are preserved?" I ask.

"Millions, billions. Literally billions are out there," he says. "They are not all exposed on the surface because you just see little patches, and then the track-bearing layer goes back under the overburden or under a mountain."

A track-bearing layer was once a thin layer of soft mud, usually around the edge of a body of water, where dinosaurs had come to drink, and others, predators, had come to feed on the thirsty. After the track-makers left and the water level began to drop, their footprints were baked by the sun and were eventually covered by a new layer of sediment. A site can sometimes yield multiple layers of tracks preserving the evidence of thousands of dinosaurs.

"One of the exciting things that we've discovered in the last few years," Lockley says, "is that several track sites are all interconnected, and that a particular layer of strata has tracks wherever you find the surface exposed. We've been calling these mega–track sites. We've got one up and down the front range we call the Dinosaur Freeway, which goes from Boulder down almost to eastern New Mexico. We can make a map of the tracks in this particular layer of strata and go to those areas we know are exposed and find more tracks.

"It basically used to be a coastal plain, like the Gulf of Mexico today. There was a seaway which came up into this area and created this huge shoreline for hundreds of miles, and there's tracks all along it. This is not unique. We have found several examples of these mega–track sites—we start studying them and we find they get bigger and bigger. This one, from the middle of the Cretaceous, is eighty thousand square kilometers [30,888 sq mi].

"I'm talking about an area the size of a couple of states. This layer goes down a mile [1.6 k] under the surface below Denver. It dips off the Front Range to the east, and then levels out in what we call the Denver Basin. We did some calculations based on the tracks we see at the surface; if you have one track per square meter [1.2 sq yd]—that's fairly typical of some of the sites we've mapped—then you've got one million tracks per square kilometer [0.39 sq mi]; if you have eighty thousand square kilometers [30,888

sq mi], then you've got eighty billion tracks. If you commute to Denver from Boulder and back every day and your car is two meters [2.2 yd] wide, you drive over a couple hundred thousand dinosaur tracks every day. The numbers are mind-boggling, but basically it means that you had a coastal plain that was just trampled for a period of a million years or so. We have other mega–track sites . . . one that's Late Jurassic."

THE LATE JURASSIC SANTA FE TRAIL

The dinosaur tracks fell under the jurisdiction of the U.S. military, which, at the time, used a stretch along the Purgatoire River to test the maneuverability of high-tech military equipment like M-1 tanks and Apache attack helicopters. Martin arranged special permission for us to go there. We signed in at headquarters, filled out another batch of forms to allow us access to study the tracks while holding the army blameless should we be run over by a tank in the middle of the night while we camped. Tank tracks crisscrossed the barren terrain in every direction as we drove down into the valley to visit the longest dinosaur trackways in the world.

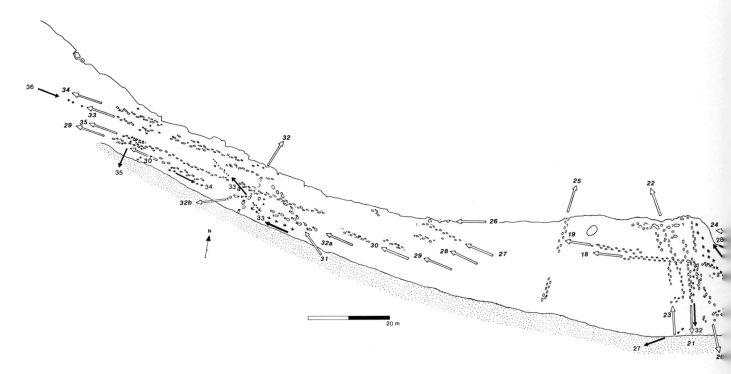

20 m

When we arrive Martin tells us, "The Purgatoire site is Late Jurassic in age, which is the Morrison Formation. It was found by Roland T. Bird of the American Museum of Natural History, Barnum Brown's right-hand man, in 1938. You can follow this layer for almost a quarter of a mile [0.4 k]. All the tracks in it fall into two types: either *Brontosaurus* tracks or these three-toed tracks, which we consider most probably made by an *Allosaurus* type of theropod. As far as the brontosaur tracks are concerned, there is a real strong tendency

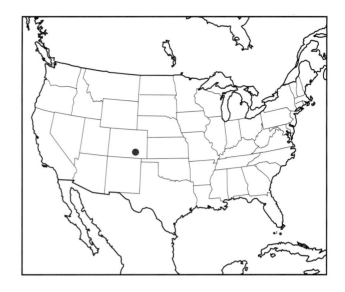

ABOVE: The Purgatoire River in southeastern Colorado
LEFT: Over half the tracks along the Purgatoire River are those of theropods, perhaps cruising the shoreline for herbivores that have come to drink.
BOTTOM: A partial view of Lockley's Purgatoire River trackway map, the world's longest continuous mapped trackway, which illustrates a total of about 1,300 tracks representing nearly one hundred trackways.

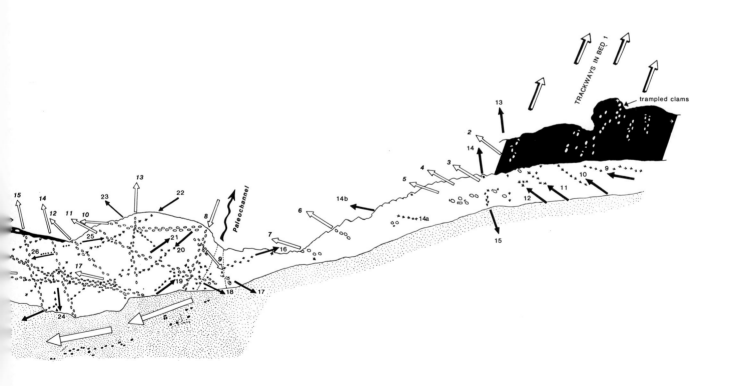

Martin Lockley crouches in the hole of a giant sauropod
footprint as he prepares to make a cast.

for the majority of them to be heading west. We've got about twelve parallel trackways that are all heading west, like the Late Jurassic Santa Fe Trail or something. It's the longest trackway in the world—about two hundred meters [660 ft]. These *Brontosaurus* tracks are beautifully preserved, and they have very, very regular spacing, almost like birds flying in formation. All five of those animals are exactly the same size within a fraction of an inch. Their footprints are all the same

size. They were all subadults, about half the size of a full-grown brontosaur. It's very, very compelling evidence for a gregarious group of *Brontosaurus* moving west. We have quite a lot of very good evidence that they are on lakeshore deposits and that they were moving along a shoreline.

"The layer just below the one with the gregarious group is trampled. We found a series of about two dozen clams, and when we started looking at them closely, we realized that none of them were in the normal position of clams in the substrate. They were all pushed to one side, and some of the shells were actually cracked and crushed in the bottom of the footprints. When the brontosaurs went stomping through here, this clam population was just decimated. In another area we found the clams had not been disturbed. They were in the normal position that clams grow or live in. This may be coincidental, but they were much larger. They lived to old age. These other ones had been snuffed out rather early in life. They were definitely trampled.

"One of the problems with bones is that big animals very often are overrepresented and you can't find the small ones because their bones are washed away. This bias in the fossil record has been debated for a long time, and nobody has tried, until recently, to consider that the tracks may add another dimension. In many ways, tracks are probably a little more reliable than bones in a general sense because there is no reason for the footprints of the small animals to be selectively removed. There can be bias toward large tracks, too, but in theory it's a little bit less likely to happen.

"Almost all tracks were originally made on a relatively flat surface, like a beach or a mud flat. I look for these surfaces where they've been exhumed again. The more surface you see, the more you will improve your chances. Quite simply, all the world's largest track sites are sites where you have big surfaces exposed.

"Tracking got a kind of bum rap in the nineteenth century . . . they couldn't tell the difference between birds and dinosaurs. Nobody's really taken this study very seriously, so the science is very immature. We've been able to make very rapid strides in the last ten years to make tracking a science. A lot of questions about social behavior and speeds can be addressed by

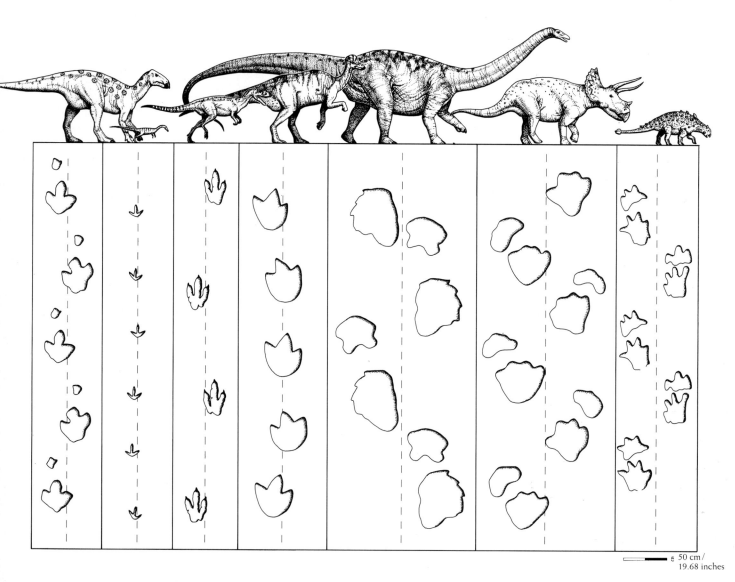

Iguanodon   Coelurus   Ceratosaurus   Corythosaurus   Apatosaurus   Triceratops   Euoplocephalus

50 cm /
19.68 inches

Artwork showing dinosaur track varieties
Illustration by Pat Redman after research by Lockley

tracks. The whole mega–track site phenomenon is an unprecedented discovery. The science has matured to the point where we can predict where to find tracks. We can go out and find five track sites today, and people go, 'Wow, that's amazing . . . we didn't think they were out there.' It's not magic—it's science."

THE CALLINGS OF FATHER GIUSEPPE LEONARDI

There are no dinosaurs in Venice—but very fine wine.

—postcard from Father Giuseppe Leonardi

The first thing I noticed when I walked onto the third floor of the Venetian home was the elephant. It was not a big elephant; in fact, Father Leonardi explained rather apologetically that it was only a dwarf elephant from Sicily—Pleistocene in age—just a million years old. Just. Nevertheless, it was an imposing and fitting hall piece for Leonardi's family, who have been scientists since the fifteenth century. Leonardi's father, Piero, age eighty-six, is a tracker himself and a world-renowned geologist. He picked up the elephant bones on one of his numerous expeditions through the rugged landscape of southern Italy. Piero Leonardi shook hands with his visitors, gave a polite nod, and shuffled back into his study, where he was busy putting the finishing touches on a monograph of the Dolomites— one of the more dramatic geological regions of the Italian Alps. When Leonardi was a boy of ten, he frequently accompanied his father to the Alps to look for the ancient trackways of extinct animals. From the tracks he learned to determine an extinct animal's physical proportions, its gait, speed, and social behavior; if the animal traveled in organized groups; if it was a herbivore or a carnivore that preferred hunting in packs.

Priest and dinosaur tracker Giuseppe Leonardi, who understands some thirty different languages, has crossed piranha-infested streams and flooded rivers and has been robbed by bandits three times in his unstoppable quest to discover dinosaur tracks.

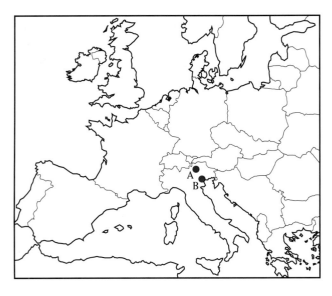

ABOVE: Dolomites in the Italian Alps (A), and Venice (B)
BELOW: Araraquara (A), São Paulo (B), São Carlos (C), and
Rio de Janeiro (D), Brazil.

In the footsteps of his father, Leonardi learned how to track an ancient animal's trail, but then he also became concerned with the path of human souls, and later became a priest. While studying for the clergy at Rome University, he pursued a Ph.D. in paleontology. His first assignment after ordination was to teach the ways of God to a people in the hinterlands of Brazil who were suffering persecution under a military dictatorship. It was dangerous work, as the military perceived the Church as a potential rival.

Father Leonardi recalls, "I had the impression that research, for me, was dead. I gave a friend almost all my books and osteological [animal bone] collections. I gave away my geological instruments. I went to Brazil just with my hammer, a compass, an altimeter, and a small collection of publications on tracks. If God wanted me to continue work, I believed He would give me a sign."

Brazil, though nearly as large as the United States, had only one known fossilized trackway. It was found by the Brazilian geologist Joviano Pacheco in 1913 in the quarried slab of a sidewalk of São Carlos. In July of 1976 Leonardi made a pilgrimage to São Paulo, where this flagstone slab was on display in a small museum. While on a long bus trip back, Leonardi developed a severe toothache and got off in the town of Araraquara to find a dentist.

"I was almost dead from the pain," Father Leonardi recalls. "I began walking the streets searching for a dentist and I began noticing the sidewalk. I noticed it was the same kind of stone as in the museum in São Paulo. Then I began discovering trackways in the flagstones and I forgot the dentist! I was hopping in the street for joy! I began walking on my hands and knees and studying more trackways. I began working and searching for more and more of them all over the city."

Leonardi quickly found the source of the tracks, a quarry near Araraquara that was excavating flagstones from a large fossilized sand dune. Back in the Early Jurassic, nearly the whole south of Brazil was covered by a desert—it is very unusual for footprints to be preserved in sand.

"I visited some present-day deserts to study the animals for comparison," Leonardi says, "and I saw that at night, many times, there is dew on the sand. The

Giuseppe Leonardi with tracks embedded in the quarried
sidewalk slabs of a cemetery in São Carlos, Brazil,
containing the tracks of numerous species of dinosaurs.
Photo by Dr. William A. S. Sarjeant

sand is a little wet, and during the evening the night-dwelling animals had left their footprints."

Along these ancient fissures, representing quite literally the difference between night and day, the quarry operators easily split apart the fossilized dune and break it up to be sold for sidewalks and the façades of buildings. When Leonardi searched the town, he began discovering footprints everywhere—on the outside of the airport terminal, the botanical gardens, the zoo. In a nearby town Father Leonardi received a divine stamp of approval when he saw dinosaur trackways on the flagstones in the cathedral.

No one in Araraquara, a town of 400,000 people, had ever noticed the dinosaur tracks except the quarry owners, José Grosso and his son Oswaldo. To them, a flagstone slab with trackways merely made the stone

defective and brought down its price. They happily put aside all the imperfect slabs for Leonardi, so that he could cart them away to the National Geological Museum in Rio. But for Leonardi some of the more important slabs still remained in the sidewalks of Araraquara.

Leonardi calculated that the city of Araraquara had 108 linear kilometers (67 mi) of sidewalks, the surface of which would cover three-quarters of a square kilometer—an incredibly large exposure.

"Generally when you collect a fossil," Father Leonardi explains, "you log in your field notebook where you found it, 'hill 233 near the canyon.' You describe the landscape of the place or you put it in the map. But for the trackways of Araraquara, I put in a town plan and put the number of the street and the number of the house marking the fossil."

It took several years to mark all the town's trackways. Ultimately he selected some sixty slabs to be placed in the National Museum in Rio de Janeiro. But the town's dignitaries were not amused.

"I spoke to the mayor about taking the tracks to the museum," Leonardi laments, "but he ridiculed me, as did his successor—they thought I was completely mad or perhaps crazy. They did not believe them to be dinosaur tracks, and they did not permit me to take away the slabs."

Father Giuseppe Leonardi then confesses.

"I began stealing slabs by night," he tells me. "One time I went with a colleague of mine with an official car of the university, and with the geology hammer and chisel, and we took away some splendid slabs. The more important of them. To make replacements, I traced each slab's outline on a newspaper, careful to mark the paper with the house number. It was very embarrassing work, kneeling on the sidewalk with paper and a scissors cutting the newspaper, so I decided to do it during Carnival. You can do all the crazy things you want during Carnival! People think you're drunk. Or joking."

His friends at the quarry, the Grossos, used his tracings to cut an exact replacement of each trackway stone.

"One time I knelt on the sidewalk and began taking away one very important slab. Some people asked

Giuseppe Leonardi removing one of the many flagstone
slabs containing dinosaur trackways in the sidewalks
of Araraquara, Brazil.
Photo by Dr. William A. S. Sarjeant

me, 'What are you doing here?' I said, 'This one is
defective. See?' and I showed them the footprints—
the imperfections. 'And I will put one back in—a
smooth one.' So they went away and I took away the
splendid slab."

I ask Father Leonardi whether he was ever bothered by the commandment "Thou shalt not steal."

"Ah, no problem," he says with a dismissing wave
of his hand. "One flagstone can have the value of a
half a dollar or so. And because I put another slab in
the same place, it was okay. I think that science has a
right of its own. And I was not stealing for myself, but
for science, for the community."

Upon the election of a new mayor, Leonardi
asked scientists from around the world to petition
the town's fathers, asking for the selected sidewalks
of Araraquara to be preserved in a suitable archive.

The new mayor buckled under Leonardi's worldwide
consortium of pressure and even enlisted a road crew
to help him excavate the fossil trackways from the
town's sidewalks.

The trackways are now in several Brazilian museums. In all, Leonardi identified twenty-five species,
among them eight dinosaurs, nine mammals, and eight
rare therapsids.

Father Leonardi resumed his paleontological work
in Brazil with renewed fervency. In the next sixteen
years he led eighty-six expeditions and discovered
more than six hundred dinosaur trackways in sites all
over South America. The perils were immense. Crossing piranha-infested waters. Being attacked by bandits.
Fording flooded rivers. The hardships leave him un-
deterred: Leonardi feels his work is divinely inspired.

"I left Italy as a missionary thinking that I would
not have any more possibility to research, and it was a
big sacrifice for me. I left many things to do missionary work: my father, my mother, my country, my language, and my friends and research. I went and now
God is giving me the centuple—a hundred times more
than I ever had.

"One time I found a splendid footprint of a Devonian amphibian—one of the more ancient footprints of a walking vertebrate in the world. It was really a lucky strike. If I stayed in Italy, I would not have discovered this; I think that God was thinking of me in Devonian. God was there when that big amphibian was leaving its footprint in the mud flat of the Upper Devonian, putting that footprint aside for me so that three hundred eighty million years later I would find it. Small things like that show you the love of God."

After several years Father Leonardi was asked to return to Venice to become the superior-general of Cavanis Institute, a high school run by a religious order. Again he left his friends, his research, and his adopted country to follow the calling of his church.

> The scene that opened from the edge of the
> pit was mountainous, and such a desolation that
> every eye would shun the sight of it: a ruin like the
> Slides of Mark near Trent.*
>
> — Dante Alighieri, *Inferno*

Such was the exiled Italian poet's vision of the staircase to hell when he wrote these words in the fourteenth century. Father Leonardi and I are standing on the ruins of Dante's castle, as it is now called, and looking at the "Slides of Mark near Trent" that inspired the poet's verse. Whenever John and I tour with Leonardi, we are blessed with his ability to recount two histories, earth's and man's.

Leonardi explains that 210 million years ago this area high in the Italian Alps was an island floating on a continental plate in an ancient sea called the Tethys Ocean. The island was tropical, more or less the size and shape of the island of Andros in the Bahamas, and its inhabitants were dinosaurs. Along its beaches was a muddy carbonate platform in which the dinosaurs left their footprints. Eras came and went as the tracks turned to stone, the dinosaurs died out, and the island, adrift on its plate, bumped up against Italy and was pushed, lifted, and fused into an upstart mountain range that became the Alps. As the plate took millions of years to trudge up the mountain, the tropical island

*John Ciardi's translation (New York: Penguin Books, Mentor Books, 1954).

had tipped to a thirty-degree incline and clung precariously to its side. In the late ninth century an earthquake struck the Alps, shaking loose a large part of the island. This voluminous mass cascaded down the mountain to the valley below, crushing a town and causing the Adige River to reroute itself several miles around the natural catastrophe.

In the fourteenth century a Florentine poet, Dante Alighieri, backed the wrong political party and, to escape being burned alive, exiled himself to the Alps. There a sympathizer put him up in a castle on a prime piece of real estate overlooking the geological remains of the legendary landslide. Dante, now sufficiently inspired by his own tortured existence, began to write his masterpiece, *Divine Comedy*, which includes *Inferno*. When Dante needed an earthly reference for his stairway to the pits of hell, he chose the landscape outside his window.

A few years ago, not far from Dante's castle in the vast acreage above the valley where the island lost its footing on the mountainside, a man was on an evening promenade with his dog, not at all aware he was also strolling on a beach of a once-tropical island inhabited by dinosaurs. He noticed some strange tracks in the stone and reported them to the local museum curator. By an ill stroke of luck he was preceded by a woman who reported she had found a set of fossilized dog breasts. The museum curator, attributing the sudden preponderance of discoveries to the phase of the moon, politely dismissed both visitors. A year later the curator scheduled a show of dinosaurs from China and was slightly embarrassed that Italy had none. The curator called in Father Leonardi to investigate the alleged dinosaur tracks, and suddenly Italy was full of dinosaurs. He found some 130 other trackways in the area.

While Leonardi and I poke around Dante's old stomping grounds among the carnivore tracks, I ask him whether he believes there are dinosaurs in hell.

"Well, I too have this idea," says the priest. "I hope they are not in hell but in paradise. I think many times that I would like to see dinosaurs in the eternal life where there will be no time—only the eternal present. Perhaps it's a little materialistic kind of idea, but it is a hope of mine."

Ever since Darwin predicated evolution as the mechanism of Creation, the Church has been strug-

OPPOSITE: High in the Andes, geologist Ricardo Alonso of
Salta, Argentina, measures the stride of a
carnivorous dinosaur with a two-meter stick on a nearly
vertical faulted cliff. The very last dinosaurs of the
Cretaceous left behind their tracks and traces along this
ancient shoreline turned to stone.

ABOVE: An eighth-century landslide in the Alps above
Trento, Italy, inspired Dante's metaphorical stairway to
hell in the *Inferno* and later revealed dinosaur tracks,
like these being studied by Giuseppe Leonardi of Venice.
Prior to their discovery there was virtually no
evidence of dinosaurs in Italy.

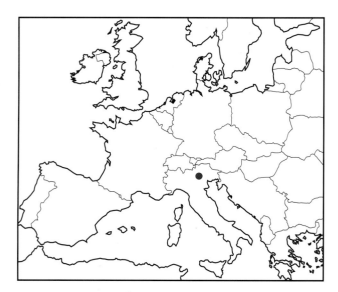

Track site south of Rovereto, Italy

WRANGLING DINOSAURS

In 1836, six years before Sir Richard Owen named the dinosaurs, Edward Hitchcock, an American clergyman and teacher, delivered a report to the *American Journal of Science* announcing what he called "the remarkable footmarks in stone in the valley of Connecticut River, which have since awakened so much interest among intelligent men." Over the course of his life Hitchcock collected over twenty thousand fossil tracks and assembled a footprint museum, still the world's largest collection, at Amherst College, where he was the president. But Hitchcock, his century's most accomplished dinosaur tracker, went to his grave believing that the footprints he collected were made by large birds.

"At the time Hitchcock was working," notes Jim Farlow, a geologist at Indiana University, "there weren't any good, complete skeletons of dinosaurs. So there was no way of knowing that some of them walked on their hind legs and had birdlike feet. The only dinosaurs they knew back then were interpreted as being gigantic alligators—or lizardlike creatures—so the closest identification Hitchcock could get was a giant bird."

The first fossil of a dinosaur was discovered in 1802 by Pliny Moody, Esq., then a boy, when he plowed up five tracks on a farm in South Hadley, Massachusetts.

"So strikingly did these tracks resemble those of birds," Hitchcock wrote, "that they were familiarly spoken of as the tracks of 'poultry,' or of 'Noah's Raven.' " William Wilson, who discovered stone tracks in flagstones being laid in the streets of Greenfield, Massachusetts, and claimed to be the first to bring them to the public's attention, said he observed them as tracks of the "Turkey Tribe." Hitchcock himself, in the mounds of scientific papers he generated, referred to them as *Ornithichnites*, or "stony bird tracks."

Jim Farlow, a well-known dinosaur tracker himself, uses the fresh footprints of living birds to help interpret dinosaur trackways. We joined him in Ft. Wayne, Indiana, for a footprint-making session with his current subject, the emu, a very large and, we were told, extremely stupid flightless Australian bird.

"They are the largest living three-toed ground-living bird," Professor Farlow explains in his office the

gling to reconcile Genesis with these discoveries. Since Father Giuseppe's feet rest in both camps, I ask him how he resolves the conflict.

"I am not a creationist," Giuseppe tells me. "Creationists think that it is possible to demonstrate by science that the world was created in six days of twenty-four hours. Now, as a biblicist, I know that the intention, not only of God but the intention of men who wrote these texts, was not to explain the scientific way of Creation. That text is poetry. It is a song and it is beautiful. But it is not a scientific demonstration. The Bible is a light on salvation—it is not a textbook for geology."

One of creation's most dynamic forms was *Tyrannosaurus*, the "tyrant king." "Why would God create a *T. rex?*" I ask Leonardi.

"Oh, that you must ask to Him," he answers with a smile. "I think that God is generous with life. He gave space to many forms, many animals, all of them important, they are splendid. They are also, like us, in the image of God. Not anthropomorphically, but because we are alive, intelligent, and beautiful.

"So *Tyrannosaurus rex* is an image of God. Because he has life, and God has life. Because he is beautiful, and God is beautiful. Because he has a meaning, and God has a meaning. In many, many ways, each animal, each star, each butterfly, each flower, are an image of God. So dinosaurs are also."

day before the session. "An ostrich is bigger than an emu, and so it is closer in size to a dinosaur, but an ostrich has a very peculiar two-toed foot, rather than a three-toed foot. So the emu is the next-best choice. Emus, however, are dumb as all get-out. Getting them to do what you want them to do can be extraordinarily difficult. I have frequently prepared beautiful surfaces for them to walk on; where if they will walk on it I will see every scale of the sole of their foot. And while I'm preparing the surface, we cannot keep them away from us. And the moment it's ready and we want them to walk on it, they won't go near it. You herd them toward it and they stop and jump over it. They can be incredibly uncooperative.

"But at least they aren't mean. They are very docile, very friendly. Unlike ostriches and rheas, they don't go into a footprint-making session with the intent of killing you. The ostrich keepers have these kind of forked metal sticks with a big end—and you go into the pen with the ostrich, and the ostrich says, 'Ah! Somebody to kill!' and starts coming at you. The handlers brace themselves holding these sticks. It's like every bad caveman movie you have ever seen."

Jim told us he was a little understaffed and deputized us to help him wrangle an emu for the footprint-making session. Early the next morning John and I showed up at the Ft. Wayne Zoo to help Farlow smooth out some freshly tilled topsoil for the emu footprint experiment. I couldn't help noticing that Farlow was wielding a large forked stick like the ones in bad caveman movies.

"Emus have a huge nail at the end of their toes," he explains. "They could take quite a hunk out of you. But they don't do it deliberately. It's manslaughter instead of first-degree murder if they get you."

"Great," John mutters under his breath, "we'll be mauled to death and the bird will get probation."

The zoo had emus, but they were too young for this study, so a large one was being imported from one of three nearby emu ranches which specialize in raising exotic birds for yuppie pets and exotic food.

The emu wrangler, his wife, and their small son arrive later in a pickup truck towing an enclosed U-Haul trailer. The sound effects issuing from within sound more like they're delivering a gorilla. As the trailer bangs and shakes behind him, the emu's handler

ABOVE: Edward Hitchcock, clergyman and professor who, in 1836, thought dinosaur tracks were made by large birds. Scientists 150 years later have come full circle and now think that birds *are* living dinosaurs. *Amherst College Special Collections and Archives* BELOW: Ft. Wayne, Indiana (A), and Amherst College in Massachusetts (B).

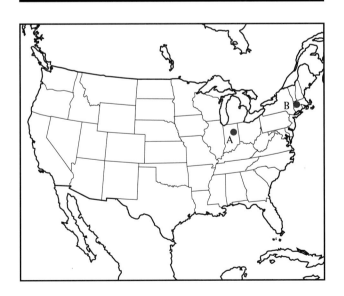

lifts up his T-shirt displaying a fresh wound streaked across his chest, inflicted just that morning when he tried to put the emu's mate in the trailer. As if on cue, the emu scratches its nail along the bottom of the trailer, causing a shrill, diabolical scraping sound to echo throughout our enclosed zoo compound.

Farlow, sensing his newly deputized help's unease, sets his caveman stick down and demonstrates with an imaginary emu before him how to subdue the large flightless bird. With his left hand he grabs its imaginary neck and with his right hand reaches between its legs and locks onto an area I take to be the dinosaur descendant's crotch, a maneuver, he explains, sure to have a sedative effect in the remote chance it should become suddenly violent. He picks up his caveman stick and approaches the trailer with John and me (in that order) taking positions directly behind him.

Farlow lifts the U-Haul gate a crack. He takes a ring of keys from his pocket and shakes them into the black hole of the rental. He whispers over his shoulder that emus like the noise. Like magic—the trailer is becalmed. Professor Farlow slowly opens the back gate, and the giant bird, like some demonic oversized poultry, stands transfixed, oblivious to everything except Farlow's jingling keys. Farlow, still jingling, begins entering the trailer talking to the beast like a detective trying make a killer surrender a gun. Then with absolutely no warning the emu, with a lightning burst of speed, shoots toward Farlow, who sidesteps the fleeing bird, leaving John and me squarely in the giant creature's escape route. I backstep over the emu wrangler's son, who unbeknownst to me has taken refuge behind me. The bird leaps over the sprawling clump of humans, his deadly foot, Exhibit A, in a foiled manslaughter attempt, whizzing by my face like a sickle.

We pick ourselves up. Behind us the emu inspects the confines of his huge compound. From the edge of the grass, the bird sniffs at the freshly tilled trackmaking arena—and concludes, judging by his future behavior, to avoid it like a minefield.

"I think that bird's smarter than all of us put together," John surmises.

"And better armed," I add. "Did you see those feet?"

For two hours we tried unsuccessfully to herd the bird through our footprint arena, but the best print

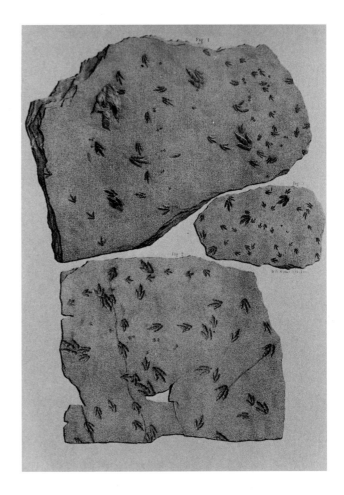

ABOVE: "Stony bird tracks," described by Edward Hitchcock in the nineteenth century.
OPPOSITE TOP: Appleton Cabinet, Amherst College, Massachusetts.
OPPOSITE BOTTOM: Hitchcock's tracks in Appleton Cabinet
Photos courtesy of Amherst College Special Collections and Archives

ABOVE: The lethal foot of an emu
OPPOSITE: Jim Farlow of Indiana University uses
the tracks of dinosaur descendants, such as the emu,
for comparison with their more exalted ancestors,
the dinosaurs.

Jim chases a dinosaur relative to acquire data for his footprint studies.

was taken from a bit of muddy area along the fence not intended for track-making.

Breathless and sweating profusely, Farlow begins mixing plaster to pour in the emu-track depression. After expertly avoiding us all morning, the emu suddenly makes a spirited charge in our direction. We scatter and the bird obliterates the footprint Farlow was preparing. Farlow explains that the next-best track, while regrettably inferior, is adequate for science. He pours some plaster into the footprint, and while it dries we watch Farlow round up the emu.

The emu and Farlow, both exhausted now, circle each other like weary fighters in their tenth round. Maintaining eye contact, Farlow reaches into his pocket and jangles his magic keys. Again, the bird is hypnotized, allowing Farlow to apply his patented emu wrestling hold. The subdued dinosaurian poultry stares dumbly into space as Farlow gently picks him up and sets him back in the U-Haul.

Acquiring a worthy footprint under the best of conditions is remarkably difficult. A fossilized dinosaur footprint surviving millions of years now seemed a miraculous event.

"Any particular footprint has a very, very poor chance of becoming a fossil," Farlow later explains back at his office, "and you can think of all the reasons why. If it dries out, it might crumble apart. Or if it is made under water, a wave current may swash it away. Or another animal may step on it. However, when you consider how many footprints one animal may make in a lifetime, and you consider the number of animals alive at any one time and the gazillions of tracks being made, the odds that some of them will become fossils then become very, very good. Almost very likely.

"There are a couple different ways footprints are preserved. If the mud that the animal is walking through is of the right composition, and it bakes in the sun, they can become quite hard. And then if that footprint is filled by sediments, preferably sediments of a different kind than the ones that the animal walked in . . . the chances of preservation can be quite good."

Farlow, now forty-three, explains that the pivotal event that led him into his career was seeing the dinosaur sequence in Walt Disney's *Fantasia* when he was five. But the event that led him into tracking came later and quite by chance.

"About ten years ago I got a call from a friend I had known in college who had married into a Texas family with a ranch a couple hours west of Glen Rose. A flash flood had stripped away a lot of rock and exposed a footprint site out there. He knew I was interested in dinosaurs and wanted to know if I would come work on the tracks. So I did. That's where I found the footprints of the running dinosaurs. They were, to my knowledge, and still are, the fastest dinosaurs on record."

I ask him how big the dinosaurs' tracks were, and Farlow opens a drawer of a filing cabinet next to him and pulls out a plaster mold of a three-toed meat-eating dinosaur.

I measure it at about fifteen inches (38 cm) long. "That's a big guy," I say.

"Well, for New England, it's good-sized, but for Texas, no. Now, in the case of this guy, what we are dealing with is not a sixteen-foot [5 m] stride, but a sixteen-foot [5 m] pace—right to left."

Farlow calculates the top speed of this Jurassic carnivore to be about 42.8 kilometers (26.5 mi) per hour, faster than even the fastest of humans.

"After I had done that brief paper, I thought I would go to the technical literature and find out that somebody had studied footprints at Glen Rose and other places in Texas, and that would tell me what I should call these footprints. I discovered that there was very little that had been done, so I realized that if I was going to do any kind of systematic description of these footprints, I was going to have to start from scratch and basically do it myself.

"One of the problems you run into in Texas and in footprints from other localities like Connecticut is you find two different footprints. And you wonder, *Well, they are sort of alike. Are they likely to be made by the same kind of animal or by different kinds of animals?* We don't have any feel for that at all right now. That's what led me to the study of emus and other birds. I can photograph them while they are making the tracks and see what they are doing. I can see the features of the surface they're walking on, or whether they are walking or running or turning or speeding up or slowing down, how this affects the shape of the footprints, if at all. Once I have finished the emu work, I hope to have a handle on the matter of how much variability you get within footprints made by animals of the same or of different species."

For years the bone hunters considered tracking an inexact science, a discipline with too many unknowns to be taken seriously. Many trackers, such as Farlow, now use trackways to determine data you can't get from bones.

"You can learn a great deal about how the animal is moving," Farlow explains. "How it is carrying weight on its feet. It can confirm, for example, anatomical interpretations that dinosaurs were walking in an erect fashion and not sprawling like a lizard.

"Now in dinosaurs, at least in the hind limb, they seem to have been very much more erect, like the emu trackway. They are walking by putting one foot almost directly in front of the other. Paleontologists had always suspected that, because of the way a dinosaur femur is constructed. But the trackways prove it.

"You can also learn something about whether these animals were solitary or gregarious in their habits. For the sauropods, which are big brontosaurs, most of the time when you find trackways of these dinosaurs you find more than one, and they tend to be going in the same direction. And that sort of suggests that these guys were rather social in their behavior, that they moved around in huge herds.

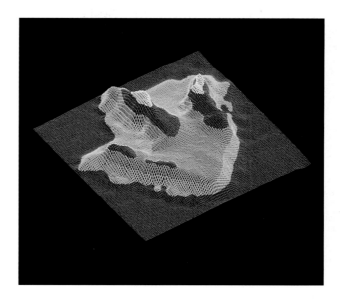

ABOVE: Farlow's computer-mapped image of a Texas therapod track.
BELOW: Site of the fastest known dinosaurs—on a ranch twenty-two miles (3.5 k) northwest of Junction, Texas.

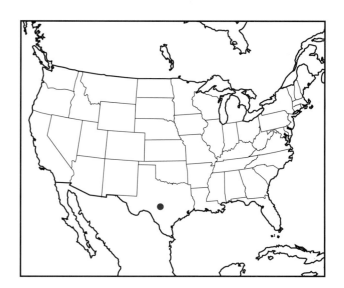

"There is evidence that suggests at least some carnivorous dinosaurs moved in groups, too—packs perhaps. I've even speculated that during the Jurassic Period some of the big meat-eating dinosaurs like *Allosaurus* might have ganged up on some of the big sauropods in packs to kill them."

"Can you tell what animal made what track?" I ask, admiring the giant foot in his hands.

"To a certain extent you can. You can tell if you have got a three-toed track that has got long, skinny toes and evidence of sharp claws at the end that it is probably a meat-eating dinosaur. Whereas one that has

Glen Rose, Texas, theropod footprint with author's (size 13) foot.

got rather short, stubby toes is probably an ornithopod—a plant-eater. It is very hard to say for sure exactly which theropod made a theropod footprint, or which ornithopod. There it's a matter of looking at what you have in the skeletal record from that time and place to see if you can find something that matches it. Something that has a foot skeleton of the right size and shape to have made that track. But you can't be a hundred percent certain."

One thing that puzzles dinosaur trackers is the preponderance of meat-eating-dinosaur tracks. In the fossil record there are simply a lot more meat-eating-dinosaur tracks around than herbivore tracks.

"That sort of boggles the imagination," Farlow explains. "It's like the mythical Chinese village where everybody makes a living doing everybody else's laun-

dry. It just would not work very well in a realistic, ecological sense to have more carnivores than herbivores. We know from skeletal faunas, and in modern ecological studies, that plant-eaters are generally far more abundant than carnivores. So a couple things may be going on here.

"One is maybe we are totally wrong in our identification of these footprints after all! Which I don't think is what is happening but I suppose is a possibility that can't be ignored. I suspect that a more likely explanation is that these meat-eating dinosaurs are simply more active than the plant-eaters were. And there is some evidence for this kind of thing from studies of modern mammals. Carnivores tend to have much larger home-range sizes than do herbivores. And they tend to walk a greater distance in a day than do the herbivores. And if you think about it, the reason for this is not that mysterious. If you are a plant-eating animal, you probably don't have to go very far to find more plants. Whereas if you are a carnivore that is feeding on large prey, this prey is fairly widely dispersed. If the same was true of the dinosaurs, you could have a situation where perhaps the meat-eaters, because they are more active, are making more footprints in a day than are the plant-eaters. It's a kind of behavioral interpretation.

"And still another possibility, I suppose, is that we are only seeing a small subset of the environments that these guys were customarily occupying. Maybe the meat-eaters were preferentially patrolling these muddy substrates because they knew that eventually those plant-eating guys had to come get a drink.

"In a nutshell, I think that a large carnivorous animal has two conflicting problems that it has to solve. One, if you have too many in an area, they are going to eat out their prey. As a result, they tend to be restricted to fairly low population densities. On the other hand, if you have too few of them in an area, then they have too large a geographic range to maintain, so they get stuck in kind of a delicate balancing act. I think that for various reasons mammals have not been able to balance this act as well as dinosaurs—possibly because dinosaur populations turned over more quickly or because they had lower food requirements than you will find in, say, a hypothetical tyrannosaur-sized tiger."

Floyd the Navajo Tracker

We hired Floyd Stevens, a Navajo Indian, to take us to the remote dinosaur trackway site because he was the only one we could find who knew the way. Only later did we find out that he was also a skilled tracker who helped his neighbors find lost livestock in the vast, fenceless expanse of Indian country. Through his uncanny powers of observation of the tracks of living animals, Floyd uncovered the fossilized trail of one of the fastest dinosaurs in the world.

The site is in a remote area of the Navajo Reservation, north of Flagstaff, Arizona. Barnum Brown of the American Museum had originally discovered the site in the 1930s, but the location was lost when a flood buried the track site in a thick coat of mud. Only in the last few years has it been rediscovered.

One hundred eighty million years ago, dozens of meat-eating dinosaurs had crossed this once-gloppy mudflat, leaving behind footprints that have since turned to stone. The exposed area of tracks was about the size of a tennis court, but many more lay underneath the surrounding sediment.

To Floyd, the trackway site was like a giant puzzle, and he wasn't satisfied until he had tracked the path of every dinosaur. Some dinosaurs' tracks were easy to follow, like one probably made by a *Dilophosaurus*, a thousand-pound (454 kg) predator who left a path of eighteen-inch (46 cm) footprints so straight and purposeful that it looked like it could have been laid down with a surveyor's transit. But smaller and much lighter dinosaurs left weaker impressions and were more difficult to track. When a dinosaur seemed to have vanished into a crowd of others, Floyd would pursue the beast relentlessly through the maze until the track finally plunged under a blanket of sediment.

One trackway seemed to stump Floyd. There were three ten-inch (25.4 cm) footprints in a row. The right and left footprints appeared to be eight feet (244 cm) apart—an impossibly long stride for such a small creature. Floyd was crawling from one track to the other, studying each track as if he were scrutinizing the small type in a contract.

Standing over Floyd, I suggest the only logical conclusion: Some of the animal's intermediate tracks must have been obliterated.

Floyd looks up at me and says simply, "It's running."

I reason with Floyd. "It can't be running. Those tracks are eight feet apart. Do you have any idea how fast this animal would have to be going?"

He looked from track to track and nodded his head in agreement. "Fast—very fast. But it is running."

There are only a few recorded trackways of running dinosaurs in the world. To some paleontologists, this affirms their notion that dinosaurs were cold-blooded and therefore slow.

With a tape measure Floyd measures eight feet (244 cm) straight back from the last visible track and draws an X on a layer of sediment still covering the track-bearing layer. He walks to his truck and retrieves a jackhammer. He begins carefully chipping away around the X marked on the overburden. When the slab is worked loose, he gets down on his hands and knees and with a surgeon's care eases the stone over. Underneath is a perfect matching three-toed impression. He gives me one of those I-told-you-so smiles.

The dinosaur was running! Floyd the Navajo tracker had uncovered a Jurassic athlete whose cruising speed we calculated at 23.3 kilometers (14.5 mi) per hour, as fast as a warm-blooded Olympian long-distance runner and one of the fastest dinosaurs on record.

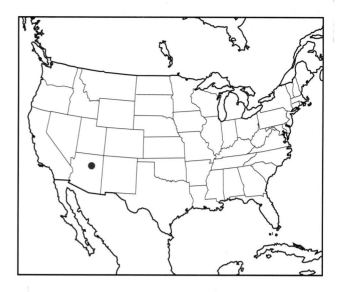

ABOVE: Track site near Cameron, Arizona
OPPOSITE: Evidence of running dinosaurs is extremely rare in the fossil record, but our Indian guide, Floyd Stevens, tracked one down in a remote area of the Navajo Reservation. Floyd's son Jason tries unsuccessfully to match the ancient predator's stride, which was as fast as an Olympian athlete's.

THE CRETACEOUS

FROM FLOWERS TO *T. REX*: IT WAS THE BEST OF TIMES, IT WAS THE WORST OF TIMES

In the Cretaceous, Mother Nature debuted her crowning terror, the biggest, baddest land predator of all time: *Tyrannosaurus rex*, "the lizard king," nature's terminator. And the essential ingredient for its wicked reign of terror may have been flowers.

Angiosperms, the technical name for flowering plants, proliferated in the Cretaceous, an event that many paleontologists think led to the success of large dinosaur herds, which would have had a quickly rejuvenating food source. Fossil evidence indicates that the ceratopsians (the horned dinosaurs) and the hadrosaurs (duckbill dinosaurs), two of the most evolved plant chewers of all times, roamed in tremendous herds. Hadrosaurs, wielding up to twelve hundred teeth at a time, are regarded as Cretaceous Cuisinarts. A herd of these voracious masticators could have turned a forest into a salad in short order. Their equally endowed contemporaries, the horned dinosaurs, with gigantic beaks, also roamed in large herds and are imagined to have clear-cut their way through forests as they migrated to their next supper.

By extension, the viability of flowering plants, followed by the success of the giant herds, inevitably led to the ultimate success for the king of the food chain—*T. rex*, who would now have a vast quantity of Cretaceous vegetarians for prey.

The Cretaceous marked the debut of the biggest predators ever found, but in the end it was the little guys, turtles, frogs, and the fabulous furballs—the mammals—that survived the great extinction.

Bob Bakker told me one afternoon as he dug out the ribs of a giant predator, "Extinctions favor the small—the meek shall inherit the earth."

And Edward O. Wilson, the famous Harvard biologist, seems to take Bakker's statement one step further:

> In terms of diversity or numbers of living individuals the dominant social creature since the Cretaceous has been the ant. Presently there are nine thousand species of ants numbering something like a million-billion individuals. Even by weight they rival the combined weight of humanity.
>
> —Edward O. Wilson, personal communication

Centrosaurus stampede
© Douglas Henderson, from *Dinosaurs, A Global View*

THE STRANGE MUTATIONS OF THE EARLY CRETACEOUS

BONE APART

During the Jurassic, South America had unzipped from Africa and Antarctica and was floating by itself in the ocean for a large part of the Mesozoic, allowing the fauna there to evolve in its own strange and peculiar way.

On the last day of excavating a carnivorous dinosaur, Argentinean researchers found its head. "My God, it's a bull!" said one of them, realizing the dinosaur had horns, and at the same time unwittingly giving the beast its name. *Carnotaurus* means "meat-eating bull" and is the only horned predator known. This beast, as well as many other of Argentina's Mesozoic monsters, was described by José Bonaparte, one of that country's most notorious paleontologists, a legend for the strange specimens he describes and famous throughout Argentina for his rough and sometimes careless treatment of valuable specimens. During a one-month visit, we saw him break more specimens than some paleontologists find in a field season.

Another bizarre Argentinean creature is *Amargasaurus*, the only known sauropod to have a sail down its

ABOVE: Two of the leading dinosaur sculptors, Stephen and Sylvia Czerkas, created this beast after making a pilgrimage to the *Carnotaurus* excavation site where they discovered huge patches of fossilized *Carnotaurus* skin impressions.
OPPOSITE: One of the strange mutations that developed when South America split off from the other continents for 65 million years was *Carnotaurus*, the "meat-eating bull," a predator that grew horns on its head. This beast, like many other strange Argentinean critters, was described by José Bonaparte.

back. This "jibbed dinosaur," as John calls it, was discovered by Guillermo Rougier, a student of José Bonaparte's. Like most of the dinosaurs José mounted in Argentina, this creature's bones rested on top of each other like a house of cards; if the bottom bone moved, the whole animal risked collapse. Preparing for a photograph, José was sliding the neck and head of the *Amargasaurus*, mounted on its rickety frame, across a table when the bottom of the frame stuck in a groove and the one-of-a-kind sauropod began to topple. John, who has a structural engineer's knack for spotting impending disaster, reached over, grabbed the metal frame, and rescued the animal from another doom. I gasped in relief. But the event, we would later realize, was merely an omen for the horrors to come.

We were eight hours' drive from the nearest town, looking for dinosaur eggs deep in the wilds of Patagonia. We had learned from Jamie Powell, a paleontologist from Tucumán, that a rancher had discovered about a half dozen perfect dinosaur eggs on his ranch south of the Río Negro. We hired José and his preparator, Raul, to take us to the location.

I sleep my first night in Patagonia stretched out between the two bucket seats of a rancher's abandoned water truck, the small of my back supported by the gearshift knob. I wake up feeling like a beetle specimen stuck with a display pin. I look down at John on the dirt. He hasn't fared much better. He is surrounded by a half dozen mangy ranch dogs who during the night had curled up with him for warmth and companionship. He will smell like them for the next three days. Bonaparte emerges from the back of the mattress-padded pickup truck and leads our small expedition party to the site.

The wind is blowing about seventy-five miles (121 k) an hour on the plains, sandblasting our faces as we hike around the corner of a butte. Trying to keep his trademark white hat fixed to his dome, José shows us a spot on the ground covered with dinosaur eggshell. "But where's the nests?" I ask.

"Right here," he says, pointing to the quarter-inch-thick shell on the ground. "They're destroyed by the weather."

I am not a paleontologist, but I had just spent the good part of a month with Jack Horner, the Montana fossil hunter, who knows more about how to find dinosaur eggs than perhaps anyone. From Horner I quickly learned, like any novice paleontologist, that what lies on the ground rarely comes from there. Erosion usually sends bone and eggshell cascading down from the side of a hill.

John and I follow a trail of eggshell up a nearby hill, and when the trail stops, we begin digging into the hill with our fingernails and a Swiss Army knife. We soon discover the bottom halves of eleven humongous eggs in a row, exactly where a mother dinosaur laid them in the Late Cretaceous. The top halves probably weren't around because the babies would have hatched out the top, leaving the bottoms intact.

For protection against predators, animals in a nesting ground will usually make their nests as close together as they can without bumping into each other. Birds at a nesting site will space themselves a wingspan apart. Dinosaurs, Jack told me, probably spaced themselves far enough apart so that their tails wouldn't swat each other when they turned.

On a hunch, I walk over about thirty feet (9 m) at the same level on the hill and see another trail of

ABOVE: Señor Don Berthe found several perfect fossilized dinosaur eggs at his ranch deep in the Late Cretaceous outcrops of Patagonia.
OPPOSITE: Guillermo Rougier, discoverer of *Amargasaurus,* a "jibbed" sauropod.

eggshell coming out of the hill. Soon we uncover the bottoms of the eggs of another nest at the same level. It was quite possible that the whole area, like Jack Horner's *Maiasaura* nesting site in Montana, was a huge ancient breeding ground. To find the perimeters of the site, I want to hike over to some badlands a few kilometers away and search for more eggs. While José retires in the back of the truck for a mid-morning siesta, we march on, leaning into the wind at about forty-five-degree angles.

At the same geographical horizon about five kilometers (3 mi) away, we find two more nests, increasing the chances that this area was once a huge dinosaur nesting ground. I can't tell what kind of dinosaur eggs we're finding because we aren't finding embryos, but the eggs are huge, each about as big as a gallon jug. We have found one nearly perfect colossal egg that we want to give to José's museum, the Argentina Museum of Natural Sciences in Buenos Aires, which has rather poor egg specimens and none on display. We march back to the truck to wake José so we can give him the honor of excavating the monstrous egg, but he wants to push on for Saturday night fun in the next big town. I remind him that we are there to take pictures.

"You should see the egg, José," Raul tells him. "It is the biggest one I have ever seen."

José suddenly has an insight. "Perhaps a picture of me excavating your egg would make good propaganda," he says eagerly.

Indeed, José whistles through his teeth when he sees the egg and claims that it is the biggest one he has seen.

The specimens José has excavated have a chewed-up look about them that I thought could be blamed on

BELOW: A nest of *Mussaurus,* "mouse lizards"—prosauropods of the Late Triassic and some of the smallest dinosaur specimens ever found—were discovered by preparator Martin Vince of the University of Tucumán in northern Argentina. The animal was described by Vince and Bonaparte.
OPPOSITE: The skull of *Mussaurus* on a fingertip

HUNTING DINOSAURS

ABOVE: Patagonia eggs
BELOW: An egg-laden *Titanosaurus australis* has scraped a shallow depression on an Upper Cretaceous lakeshore. Her claws have softened the sun-baked sand to gently catch the eggs as they exit her cloaca.

weathering. Then we witness José's excavation technique firsthand. Raul, José's preparator, and I had been using a paintbrush to whisk dirt away from the delicate egg. When José arrives, he uses a hammer and chisel. As eggshell goes flying in every direction, I carefully pick up the precious fragments, hopeful that someone can one day piece them back together. Then when José gets up to relieve himself—he steps on the pile of eggshell without apology.

Next José takes us to a Cretaceous sauropod site, discovered by an electrical lineman. Even though José has already excavated the beast, he wants to show us the site for its scenic potential. As we walk to the site I ask him if he has found the skull, for any part of a sauropod skull is a rare paleontological prize. He admits he hasn't, but when we arrive at the site we are shocked to find so much bone still laying about on the surface.

Out of José's earshot John tells me, "If this sauropod was a jigsaw puzzle, I would think someone might want a few of these pieces." John reaches down and picks up a large U-shaped piece of bone—it's the lower jaw of a sauropod! John walks over and gives it to José, who silently wraps it up with his handkerchief and puts it in his pants pocket.

Our next stop is the museum in Plaza Huincul in the Neuquén Province, where Rodolfo Coria, the museum's paleontologist, also a former student of José's, has exhumed the vertebrae of an unnamed sauropod, the largest ever found from the Cretaceous. Four huge vertebrae—each weighing several hundred pounds—of this giant rest on a low, slatted table. In their current positions they are not very photogenic, so I ask Rodolfo if we can move them so that they would be in sequence. José steps in and takes charge. He maneuvers a small crane used to lift engines out of cars. With a tattered rope Bonaparte begins tying a series of knots, none of which John or I recognize.

John, an accomplished sailor, whispers to me, "If you can't tie a good knot—tie a lot of them." I cover my face in fear of what will happen next.

As José jacks up the arm of the lift, his knots slip, causing the bone to crash on the table below. Miraculously, the bone holds together and José begins retying another jumble of knots.

"José, I can tie a knot that won't slip if you want," John offers.

ABOVE: Bonaparte scrambles a four-liter egg discovered by the authors in Patagonia.
RIGHT: Buenos Aires (A), egg site on a secluded ranch south of the Río Negro River (B), and the town of Plaza Huincul (C).

THE STRANGE MUTATIONS OF THE EARLY CRETACEOUS

José scowls, "I've been doing this for thirty years —what do you know?"

"Just trying to help," John says apologetically.

Bonaparte manages to get that bone in position without breaking it, but the next massive vertebra isn't so lucky. Like the others, it weighs several hundred pounds. When he sets the vertebrae down with the crane, I notice that a large notch at the bottom of the bone got caught between the slats of the table. As Bonaparte orders Rodolfo and Raul to push the bone into a vertical position in line with the others, I tell José, "If you move the bone upright, José—it will break—it's stuck in the table."

Bonaparte glares at me and shouts his mantra again. "I've been doing this for thirty years—what do *you* know? PUSH," he yells at his two terrified helpers, Raul and Rodolfo. They both push, and the bone breaks into at least a thousand pieces. A large piece of the bone swings from the crane's rope. I'm devastated. Tears well up inside me. I apologize profusely to Rodolfo, and then to my horror I see José sweeping the pieces of bone into a wastebasket.

"Please don't do that, José," I plead. "Let's set the bone down and try and put it back together." Bonaparte ignores me. He tries to wire a large piece of the fallen bone onto the piece still hanging from the rope, but a two-inch (5 cm) gap remains. He simply fills the gap with putty, and now the bone no longer has its original shape. I feel ill.

"Looks great!" John offers Bonaparte with cheerful sarcasm. "You've been doing this for thirty years, that must be why they call you José *Bone Apart.*"

Everyone, even José, laughs, breaking the tension in the room. Then I go outside and throw up.

Since 1976 Bonaparte had received an unbroken string of grants from the National Geographic Society. Shortly after our visit his patronage from the Society was withdrawn and funding dispersed to other South American researchers.

At the Municipal Museum in Plaza Huincul, Rodolfo Coria, foreground, the leading paleontologist in the province of Neuquén, and Raul Vacca prepare the vertebrae of an unnamed sauropod, the largest ever found from the Cretaceous.

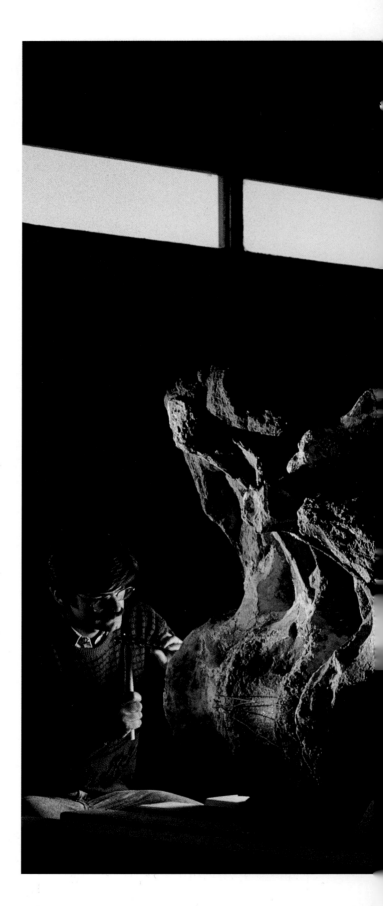

THE MEDICI OF DINOSAURS

When Giancarlo Ligabue came to the phone he greeted me like an old friend. "A dinosaur book! Fantastic! Where are you? I'll come pick you up." I was grateful for the lift; John and I were down to our last few thousand lire. Ligabue's chauffeured mahogany boat picked us up at our base camp, a cheap hotel on the outskirts of Venice, and whisked us off to his home on the Grand Canal.

I felt like I had walked into a James Bond set. Ligabue's home, designated a palace on local maps, was filled with towering canvases and artifacts from a scattering of ancient civilizations. He is one of Europe's top businessmen, but in the dinosaur world he is known as a collector and connoisseur of dinosaurs.

A man of his wealth and taste could be a patron of any of the great arts: dance, music, the visual arts. Why dinosaurs?

"Dinosaurs were my first love," he admits, pouring glasses of a fine red wine. "When I was young, as with everybody, dinosaurs represented what is a paradigm or archetype of dragons. It is inside me like all other children. Somebody gave me, as a gift, a fragment of a dinosaur egg from France, where there are a lot of dinosaur eggs. This one fragment was like a miracle to me."

This eggshell fragment became his Rosebud. Ligabue is now the president of the Ligabue Center for Study and Research, an institution that supports anthropology, archeology, human paleontology, and dinosaur paleontology. The center, occupying the entire first floor of his home, has launched over seventy-five expeditions, at least one per year dedicated to dinosaur research, principally with Philippe Taquet of the National Museum of Natural History in Paris.

His watercraft, a "Rolls-Royce" among all the others, draws stares from tourists and locals alike as it takes us down the Grand Canal to the Venice Museum of Natural History. The museum is closed to the public that day, but Ligabue is greeted by the museum staff like a returning king. I follow him into a large room named, not surprisingly, Ligabue Hall.

"This is the holotype specimen discovered and studied by Taquet," he says, entering the hall. "It was unknown before, and this is the only one in full display in the world."

The most unusual feature of this plant-eater, *Ouranosaurus*, is its remarkable, large sail-like structure that runs along its spine. I ask Ligabue, who eventually received his Ph.D. in paleontology with Taquet, if he knows the function of this unusual anatomy.

"The big spine is fantastic!" he says, walking around the specimen. "The spine is not a question of beauty or of male display, it is an external convector, like a radiator or a sun panel. The function was to increase its temperature because they were reptiles. They were not completely warm-blooded, because they used this system like many primitive reptiles."

I ask Ligabue what his role is in the expeditions.

"I lead the expeditions, particularly the more important expeditions," he says. "I am a businessman.

Venice, Italy (A), and Gadoufaoua, "land where the camel run away," east of Tahoua, Niger, in the Téneré Desert (B).

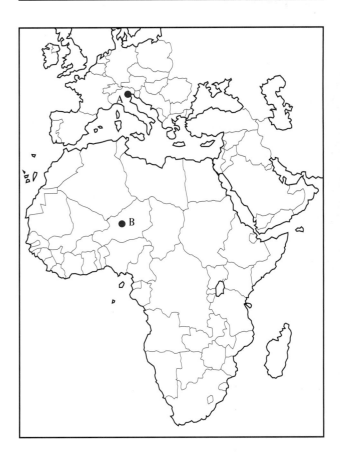

Well, instead of going yachting, I prefer to go to the desert or the forest and sleep on the ground—this is my real life. I work hard in my company, but I have more time then to devote to my research in archeology and paleontology—in particular, dinosaurs, my first love."

Giancarlo pauses and reminisces like someone remembering his first kiss. "Dinosaurs. Always dinosaurs!"

Venetian entrepreneur and dinosaur expedition leader Giancarlo Ligabue with *Ouranosaurus*, a herbivorous sail-backed dinosaur excavated from the Sahara Desert of Niger on a joint campaign with Philippe Taquet.

Angela Milner heads the Paleontology Department at the Natural History Museum in London. Here she relates the story of how she found one of the strangest dinosaur predators to come out of England.

"Back in 1983 a regular visitor to the collections brought a specimen in one day, as people often do. He came into my office and said in a somewhat embarrassed manner, 'I don't want to waste your time. I don't think this is anything interesting, but would you just look at this?' And he pulled this enormous claw out of his pocket. 'Is this interesting?' he asked. And I said yes. It was obvious straightaway that this was a claw of a very large carnivorous dinosaur.

"It turned out the claw was actually found by his father-in-law, Mr. William Walker, a plumber and amateur collector. It came from a quarry called Smokejack's Quarry, which is near Dorking, about thirty miles south of London. He'd got permission to go in the quarry one weekend. . . . He wasn't looking for dinosaurs, he was looking for other kinds of fossils, bits of plant or shells.

"He picked up this piece of rock that was sort of rugby-ball shaped and hit it with his hammer, and out

ABOVE: Reconstructed *Baryonyx* claw with nail sheath added.
RIGHT: William Walker, plumber and discoverer of *Baryonyx*, holding its claw bone.
The Natural History Museum, London

fell the claw, just like that. When he got home and stuck the bits back together, he realized the tip was missing—about the last inch. He was so upset that, the first opportunity he had, which was about three weeks later, he went back to the exact spot where he had smashed open the rock. He crawled around on the ground for about an hour until he found the missing bit. It was still there.

"The quarry dug clay for making bricks. I don't know if you've ever been in a Wealden Brick Pit in the winter. They're not very pleasant places. You slip and slide all over the place and sink in the mud. So we went back to this place where the claw had been found. We managed to detect a few little fragments of bone poking out of the ground. We put a spade into this clay and turned up several broken pieces that had obviously been crushed by the bulldozer, because there were all freshly broken pieces.

"The quarry people were very cooperative, unusually so, actually. They protected the area with some corrugated iron sheeting so the bulldozer wouldn't run over it again and let us come back in a few months when the weather was better. Once the ground dried, it was obvious there was a huge patch of iron-impregnated clay concretion lying in the ground. We couldn't be sure exactly how much of the dinosaur was there, even though we knew there was bone inside all of these nodules. We excavated the whole nodule area over the period of three weeks and brought it back here, about two tons. It took the next six years of pretty constant preparation work to get all the bone out of the rock.

"It was very, very hard indeed. It was literally a case of using fine dental tools and air mallets under a microscope, toward the end. The skull had come apart, but at least most of it was together. It would appear that the tail was lost before it was fossilized. It might have been scavenged or it might have rotted and then floated off downstream in the river or something.

"It seems the quarry was once a low-lying, swampy, marshy area. About a hundred twenty-four million years ago, southeast England was a low-lying, deltaic, estuarine area and was quite well vegetated. The Wealden has been actively and well collected for

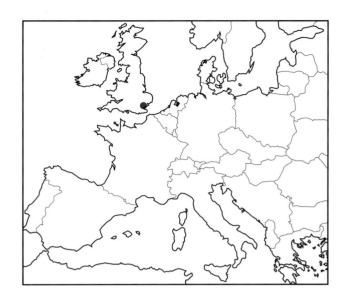

Site of Smokejack's Quarry near Dorking, England

a hundred fifty years. I think most people thought we weren't going to come up with anything new. Then all of sudden up pops this totally unique thing.

"There's no other large meat-eating dinosaur that's anything like it. It's very crocodilelike. It's not like a normal carnivorous dinosaur—when you think *Allosaurus* or *Tyrannosaurus*. Its snout is very long and low and narrow. The nostrils are set a long way back from the end of the snout, and the jaws are curved, rather S-shaped, and it really does compare very closely with a fish-eating crocodile. In fact when we first excavated some of the skull material, one or two of our colleagues said, 'Eh, you've got a crocodile, there.' Well it wasn't. Whatever its lifestyle, it obviously was doing something very different than the standard large carnivore.

"One of the unique features is its lower jaw. It's got double the number of teeth of any other carnivorous dinosaur. Each side of the lower jaw has thirty-two teeth. And all other big carnivorous dinosaurs have about sixteen or seventeen, maybe twenty teeth. The end of the snout is sort of a roseate shape which has really large teeth, just at the snout. We've come to the conclusion they were for gripping and holding something slippery, which is one of the reasons we came up with the idea that it might have fed on fish. And actually we have here one of the very few cases of finding semidigested remains. We found partly acid-etched

THE STRANGE MUTATIONS OF THE EARLY CRETACEOUS

The crocodile-like teeth of *Baryonyx*, and acid-etched
fish scales found near its stomach, indicate that it probably
fed on fish.
Illustration by Shannon Shea

fish scales and teeth inside where the ribs were etched by stomach acids.

"The only other obvious lifestyle you can evoke for it, because of the shape of its head and the fact that its nostrils are a long way back from the end of the snout, is perhaps a specialist at scavenging, which also might be the explanation for its having the remains of an *Iguanodon* in its ribs. Most large meat-eating dinosaurs have their nostrils right on the end of their snout. But in this creature they are set about four inches [10 cm] back from the end of the snout. That means you could stick your nose into somebody's body cavity and still breathe. Or you could stick your nose into water and still breathe. That's not so daft, actually. Compared to most carnivorous dinosaurs, it has got very strongly developed arms, shoulders, and hands, particularly this enormous defensive claw on the thumb—the original find. You could envision it using its arms either for fishing or clawing fish out of the water—sort of like today's grizzly bear. The one thing it obviously didn't do was just rush at things and take huge mouthfuls like the rest of the carnivorous dinosaurs.

"We decided to name it *Baryonyx walkeri*. *Baryonyx* from the Greek meaning 'heavy claw,' which is pretty self-explanatory, and the species name is *walkeri*, named after the man who discovered it in the clay pit. It was, of course, the thrill of his life to have a dinosaur named after him."

Dr. Angela Milner, director of paleontology of the Natural History Museum in London, with *Baryonyx*.

HER MAJESTY'S TERRIBLE LIZARDS FIND A HOME

Drumheller, Alberta, population seven thousand, is in a constant state of Dinomania. Just outside this small town, where the flat wheat fields drop into the badlands of the Red Deer River Valley, rests the crown jewel of paleontological museums, the Royal Tyrrell Museum. On this vast hunk of Cretaceous real estate, more dinosaurs are discovered yearly than can possibly be exhumed. It is a mecca for researchers and tourists alike, and annually attracts an average of five hundred thousand dinosaur devotees. The obvious marketing potential of a half million dino fanatics isn't lost on the commercial sector of Drumheller. From ice cream parlors to the golf course, you see dinosaurs on signs peddling products for humans.

In the early 1980s there were over three hundred Alberta dinosaurs on display all over the world, but none in Alberta. Oil and money were flowing at an all-time high, so it seemed like a good time for the province to get a proper home for Her Majesty's terrible lizards. The museum, completed in 1986, was the brainchild of Phil Currie, one of Canada's top paleontologists. During a two-year period we visited Phil several times. We went on an expedition researching dinosaur tracks

TOP: *T. rex* in the Calgary Zoo Dinosaur Park overlooks Calgary, Alberta.
ABOVE: Drumheller (A), and nearby Dinosaur Provincial Park (B), Alberta.
OPPOSITE: A *T. rex* named Black Beauty for its dark magnesium-rich bones seems to writhe in pain as a welder prepares its frame for the Ex Terra traveling dinosaur show.

in western Alberta, and spent a week in Dinosaur Provincial Park, where we watched him and his crews work on a giant *Centrosaurus* bone bed that extended to the edge of a cliff. He let us stay at his house for several weeks while we worked late nights at the museum after the crowds had gone. From his front stoop we sampled some of his privately brewed beer, planned our expeditions, talked about dinosaurs and rock and roll (his second passion), and got acquainted with some wretched Mongolian rice wine. Accompanying this modest, mild-mannered paleontologist, I found myself on the edge of cliffs looking for dinosaurs more times than I care to remember. And on July 4, 1991, John and I had the rare opportunity to participate in the worst day of Currie's life.

During a mining operation near Grand Cache, Alberta, the Smoky River Coal Company was ripping off the side of a mountain to get at a twenty-foot (6.1 m) seam of coal that runs for miles along this part of the Rocky Mountain foothills. The seam, the remains of a once-lush forest, had been tilted nearly vertical by the same geological forces that created the uplift of the Rockies. Underneath the seam one of the miners noticed some strange markings on the newly created cliff and

LEFT: A lawn ornament of a private home
near Drumheller.
BELOW: All over Drumheller, dinosaur pop culture
abounds, even at the rodeo grounds.
OPPOSITE: A 128-foot-tall (39 m) *T. rex* hot-air
balloon, owned by Thunder and Colt Balloons,
glides over Dinosaur Provincial Park.

At an Ex Terra Foundation dinosaur workshop, welders prepare steel supports for a traveling show of Chinese and Canadian dinosaurs collected from five years of expeditions in both countries.

reported them to the mine geologist, who immediately reported the find to the Tyrrell Museum.

Phil Currie went to investigate and identified the tracks as dinosaurian. They were spectacular, spread over an area perhaps two hundred feet (61 m) high and a thousand yards (914 m) long. The cliff face was like a giant aerial snapshot of an ancient Cretaceous forest floor turned on its side. You could see the root systems of ancient trees almost as if they were planted in rows for a park esplanade, and in between the trees you could see where thousands of dinosaurs had left their footprints in the mud some 90 million years ago. But of particular importance to Phil was a section of the cliff where a few lone armored dinosaurs, ankylosaurs, had wandered off by themselves. By measuring the size of an individual's foot relative to its stride, Phil could reveal the creature's dimension and speed. To get accurate measurements, Phil would have to rappel down the cliff face, which had one more characteristic particularly pertinent for us as we prepared to take photographs. The shale had a mirrorlike surface, and in the morning sun it glowed a brilliant white, like the facet of a jewel. The footprint impressions, in striking contrast, went dark. We had started early in the morning, but by the time Phil and his climbing crew had set their climbing rope, the sun had disappeared behind the cliff, making the tracks difficult to see. From an adjacent cliff I shouted over to Phil that we should return to the cliff when the light was better.

The next day we had scheduled to be at another track site that was even more spectacular. This site was discovered by a helicopter pilot ten years ago while he was dropping off fishermen in a remote spot along the Narraway River in western Alberta. He reported seeing giant carnivore tracks along another nearly vertical faulted cliff above the turbulent waters. Phil had flown in to investigate, but back then had no climbing expertise. Now, ten years later, Phil felt confident that he could reach the cliff to take the measurements.

Phil flew in first by helicopter with his team, and John and I were to return a few hours later with the helicopter to take aerial photographs while Phil measured the trackway. But something had gone wrong. Phil returned in the helicopter with his crew.

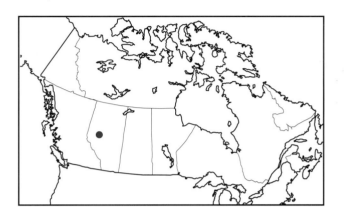

ABOVE: Smoky River Coal Company and the Narraway River site.
BELOW: John prepares gear for the Narraway River expedition.
OPPOSITE: Miners of the Smoky River Coal Company discovered a spectacular dinosaur footprint site during strip-mining operations near the foothills of the Canadian Rockies.

HER MAJESTY'S TERRIBLE LIZARDS FIND A HOME

A skull of a *Centrosaurus*, a *Triceratops* cousin

up a mining road to the top of an adjacent cliff that was to be our shooting platform, we could still hear shale crumbling down. The cliff had just fallen in the gorge! That trackway had stood for more than 30 billion days, and we lost it in a moment. Phil and his crew stood dumbfounded. We all realized that it could have been worse. If we had been just a little earlier, they would have been up on that cliff.

The next day Phil drives John and me and three of Phil's crew to the *Centrosaurus* bone bed that lies more or less in the middle of Dinosaur Provincial Park. One popular saying about the park is "If you throw your hat and it doesn't come within twenty feet of dinosaur bone, then you're not in Dinosaur Park."

I ask Phil to help me visualize the area along the Red Deer River 75 to 78 million years ago.

"During the Cretaceous this area would have been right out on the end of a delta, basically, with the sea coast off to the east, covering most of Saskatchewan and Manitoba, and of course extending all the way down to the Gulf of Mexico and all the way up to the Arctic Ocean. There would have been forests and swamps and marshes, basically everything associated with a coastal lowland, which, first of all, would be incredibly flat because you are down on the coast and this is a delta. Basically it's just an accumulation of sand and mud being pushed out into the sea, like the Mississippi Delta of today. But there still would have been a variety of environments here, depending on whether you are right beside the river or right out behind the levees. It looked probably a lot like places like northern Florida or the Gulf Coast of Texas today."

We come to the centrosaur bone bed, a platform on the edge of a cliff with bones sticking out the side.

"It's gone," Phil said when he got out of the chopper. "The cliff fell into the river. I'm sure it was the spot. We flew over it several times." My heart sank.

We drove to the mine first thing the next morning to be sure to catch the early light. When we got to the site, something looked different. I thought we had made a wrong turn. It was a big enough operation that you could easily get lost. One giant gash in the ground easily looked like another, and the mine had miles and miles of them. But this was the same place. We were sure of it. The tracks that had stood out so boldly the day before last couldn't even be seen. Was it the light? No. The tracks weren't there! As we all ran

They are getting ready to excavate. Phil and his crew have dug down to make themselves a platform of rock and dirt about six inches (15 cm) above the bone bed. We look across the valley where an ancient river continues to dig a channel now occupied by a dried-up stream. Phil explains that the bone bed continues on the adjacent cliff a half kilometer (about a third of a mile) away, and had once gone all the way across but has since been eroded by subsequent rivers and streams. At the level we now stand, a few hundred feet above the valley, there once was a river, and a few

ABOVE LEFT: (BEFORE) We were going to take a photograph of Phil Currie rappeling on this vertically faulted cliff and measuring these dinosaur tracks, but the light wasn't right by the time his climbing team set the rope. The trackways had existed for more than 30 billion days. "What was one more day?" John asked.

ABOVE RIGHT: (AFTER) We returned a day later to find that the cliff of tracks had just collapsed. To our horror, we realized that had we been a few hours earlier Phil and his crew would have been on the cliff measuring the tracks when it fell. Phil is in the center of the photo.

HER MAJESTY'S TERRIBLE LIZARDS FIND A HOME

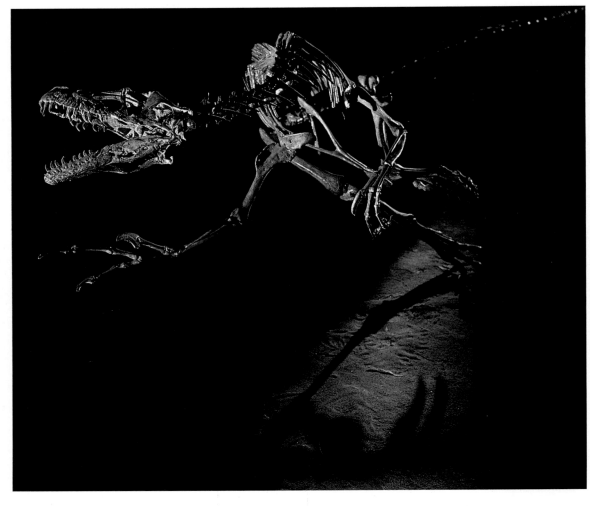

inches below our feet lie the remains of the once-mighty herd of centrosaurs, the buffalo of the Cretaceous, that died, Phil thinks, while trying to cross a river in flood.

I ask Phil how many he thinks died on that day.

"It boggles the mind. We are looking at thousands. The simplest explanation, but not the only explanation, is that a herd of these things tried to cross a river in flood. And the reason we think it was a river in flood is because the bones themselves were buried by floodwaters of a river. And so they are part of a river system, washed up to rest on sandbars and point bars and the banks of the river itself. If they died by disease or forest fire or something, I don't think this association with the river would have been that strong. They tried to swim it in a flood and they basically fouled each other up so badly that they died. Hundreds or thousands of animals. And that doesn't mean the whole herd got wiped out. Doesn't mean that many, many more didn't survive. We've certainly got lots of good examples in modern catastrophes like this with wildebeests or reindeer or whatever. And usually they all don't die. Only a percentage of them die. But because these herds are so big, it's pretty amazing what happens."

Hans Larsson, a summer intern for Phil Currie, is hanging from a rope, prospecting for bones on an edge of cliff next to the *Centrosaurus* bone bed where Phil is working. It would be a marvelous picture, I think, if I could get near him. I'm not really scared of heights—it's edges that petrify me. My father died from a fall from a roof when I was eight years old, and now I have a natural respect for high places but also a

OPPOSITE TOP: *Triceratops* at the Royal Tyrrell Museum in Drumheller, Alberta.
OPPOSITE BOTTOM: Dromaeosaur was a small but swift and nasty carnivorous dinosaur that frequented Dinosaur Provincial Park during the Cretaceous.
OVERLEAF: In Dinosaur Provincial Park, one of the most abundant resources of Cretaceous dinosaurs in the world, Phil Currie and crew dig into a bentonite knoll containing the remains of a centrosaur herd that may have died when it tried to cross a river in flood. The bone bed extends to the opposite cliffs in background.

blind passion for a good photograph. With a great deal of trepidation, I decide to join Hans on the cliff. I rappel from another rope down a gorge next to him. From the top it didn't look so far down. All I have to do is descend about forty feet (12 m) and then somehow climb horizontally over to the ledge where Hans is trying to dislodge a centrosaur toe bone from the cliff. Hans isn't unnerved by heights. I've seen him leap around the hills like a mountain goat. In fact I've seen mountain goats, from their own precarious perches, twitching their heads in curiosity as he leaped about the ledges like one of them. Hans seems comfortable hanging from a rope a couple hundred feet off the ground, even though the mountain is covered with bentonite, a kind of volcanic ash that turns the consistency of axle grease when wet, and when dry, flakes away unmercifully when you try to get a handhold or foothold.

Rappeling down to the same level as Hans was easy, but getting near him is going to be another matter. I give myself a lot of slack rope and jump over to a ledge not far away. But the ledge doesn't have a backbone and starts to give way. I wedge my hand into the bentonite to take pressure off the ledge. My knees start to shake uncontrollably as I succumb to the fear. I'm paralyzed physically and can't even call out for help. My foot starts shaking loose the precious foothold I have left. I have to find another handhold for my left hand, but I'm too afraid to lean out and find one. Then right by my nose I see a centrosaur bone sticking out of the bentonite. I work it loose with my free left hand. While holding on with my left, I shake blood back into my right hand. With the bone, I dig another, lower hole further down for my foot, and in this way I start to make a short stairway of holes over to Hans, who is working at the bone with the single-mindedness of a surgeon.

Eventually he looks over at me desperately digging handholds over to him. He leans out from his rope and calls over to me, "What are you using to dig?"

Still too scared to form words, I hold out my trembling hand with the bone.

In a low, hushed voice Hans says, "Looks like part of a centrosaur rib. Don't let Phil see you using it to dig."

HER MAJESTY'S TERRIBLE LIZARDS FIND A HOME

ABOVE: While hanging from a rope, Hans Larsson
excavates the toe bone of a centrosaur.
RIGHT: Centrosaurs crossing a river in flood
Illustration by Shannon Shea
OPPOSITE: According to Phil Currie, "The second
most common find at this bone bed is carnivore
teeth. They left their shed teeth behind like calling
cards after they fed on the washed-up carcasses of
the centrosaurs."

WHICH PLANT WILL RUST?

Kevin Aulenback is a preparator at the Royal Tyrrell Museum of Paleontology who unlocks the secrets that stones have to tell. He has drawers full of beautifully preserved three-dimensional fossil plants mounted on the heads of pins, with detail so exquisite that you can see pollen grains still stuck to leaves.

I ask him where he got them and he points out the window. "Out in the backyard," he says. In two minutes we are outside the Tyrrell Museum, roving the Cretaceous badlands of Midland Provincial Park, which was once a tropical forest.

Kevin says, "One advantage of working here is you don't have to go very far to get specimens." He kneels down and picks up a brownish rock called ironstone and points to an embedded organic pattern. "That's a pine cone off a tree like a sequoia," he explains. "It still has seeds in it. This is a good rock. It's full of plants."

The delicate plants were stuck in the middle of a rock that was essentially made of Cretaceous muds fossilized to form ironstone, which is rich in iron carbonate. It is an extremely hard brownish rock that rusts like iron and is easily visible all over the gray valley around the museum.

"The hills are really one big fossil," he says. "They're made of fossil sand, fossil bone, fossil mud, fossil plants." He explains that 74 million years ago some of the plants around here, upon being buried and fossilized, were lucky enough to be covered by silicate, a kind of natural glass, and a few of those, like lottery winners, were even luckier to be trapped in sediments that became ironstone. The ironstone encased and preserved the delicate plant, but until recently there wasn't a way to retrieve the fossil plant without great damage to the specimen. It was like having the winning lottery ticket but not being able to cash it in. Then Kevin figured out a way.

We take the specimen back to his lab, where he explains that he developed a series of acid baths that dissolve the ironstone encasement but won't damage the plant because of its unusual protective silicate coating.

"The acid will affect your hands, though," he says, as he puts on two pairs of rubber gloves. "If there's a

ABOVE: In the backyard of the Royal Tyrrell Museum preparator Kevin Aulenback probes the badlands for a piece of ironstone encasing ancient plant life. OPPOSITE: Both of these twigs are mounted on the heads of pins. On the left is a 74-million-year-old member of the juniper family (Cupressaceae) from Alberta, and on the right is its closest living relative, now found in New Guinea.

hole in one of the gloves, the acid will eat right into your hand—if you dropped it on the floor, it'd etch the concrete. You don't want to smell it, either, or it'll be the last thing you smell." Under a fume hood he adds the 15 percent dilution of hydrochloric acid to the ironstone specimen in a glass beaker and heats it to 50°C. After soaking in various hot acid baths for a day, the glass beaker is full of what looks like rusty water—all that is left of the ironstone. At the bottom of the beaker are Kevin's specimens. He strains, rinses, and then with a tweezers plucks a beautifully preserved, tiny branch out of the beaker.

Inspecting it, he says, "This plant last saw the light of day seventy-four million years ago. It's extinct now, but its closest living morphological relative is found living quite happily on a mountaintop in Papua New Guinea. I happen to have one right here," he says, pulling a sample out of a specimen file.

When I hold the two together, I find it difficult to tell which is 74 million years old. I ask him if there's much difference between the two plants. He thinks about this for a second. "If you water this one," he says, pointing to the fossil plant, "it will rust."

MONGOLIA: JOURNEY TO THE FLAMING CLIFFS IN A MOLOTOV COCKTAIL

Roy Chapman Andrews is as close as science has got to offering a character to rival Hollywood's Indiana Jones. He was a brave and daring gun-toting expedition leader who in the 1920s led a team of scientists from the American Museum of Natural History on a pioneering expedition into the innermost heart of the unexplored regions of Mongolia. Despite a hostile natural environment, roving hordes of bandits, and working in the midst of a war, they managed to make some incredible finds, among the most celebrated the first dinosaur eggs ever discovered. But the war eventually cut short their work, and the country was off-limits to Western visitors until the late 1980s.

We are flying into Mongolia with forty-two cases of equipment, not really sure if we can get it out past the customs agents, or if we do get in, if we can find conditions in which we can work. Communications in this vast landlocked country are still in the Dark Ages. All of our correspondence must go through government officials by diplomatic pouch. It takes three months to send a letter and get back a response. Our last communication from the Mongolian State Museum's director, Altan-

ABOVE: Roy Chapman Andrews, zoologist and leader of the Central Asiatic Expeditions, circa 1928.
Photo by J. B. Shackelford, courtesy of the Department of Library Services, American Museum of Natural History, New York
OPPOSITE: The forearms of *Deinocheirus* were all that survived of this colossal carnivorous dinosaur discovered in the Nemegt Valley of the Gobi Desert. The arms surround A. Perle, a Mongolian paleontologist at the Ulan Bator State Museum.

gerel Perle, came with a rare invitation necessary for our visas, but also included a note stating we should be prepared to pay thirty-five dollars for every photograph we took. It was too late to get a message back confirming whether he meant thirty-five dollars per photo or per subject we photographed—a vast difference since for any one subject I normally take maybe three or four hundred photographs. That would average out to over ten thousand dollars per specimen, well beyond our budget, but we stuff a large wad of cash in a secret compartment of John's knapsack, prepared to pay a premium for the rare privilege to photograph at least some of the collections.

All the stories we heard from people lucky enough to visit the Ulan Bator State Museum cautioned that nobody, not even visiting scientists, was allowed to take pictures of Mongolia's coveted dinosaur collections. For the duration of the flight from Beijing to Ulan Bator, the capital of Mongolia, I have a feeling of paranoid dread that upon landing, some barking Mongolian customs official will order us to kneel on the frozen tarmac, shoot us in the back of the head, and leave us for the dogs. This scenario plays out in my imagination whenever we enter a Third

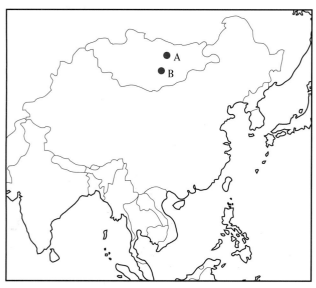

ABOVE: Close-up of the even dozen eggs
Photo by J. B. Shackelford, courtesy of the Department of Library Services, American Museum of Natural History, New York
LEFT: Ulan Bator (A), and the Flaming Cliffs of Mongolia (B).
OPPOSITE: A Mongolian child stands next to an unidentified femur of a sauropod found in the Nemegt Formation in the Gobi Desert.

World country and over half the luggage on the plane is ours. As usual, my fears are unfounded.

At the airport we are met by two cheery Mongolian officials sent by two different agencies to expedite our passage through customs. We were not expecting them, and they were not expecting each other. Maintaining polite smiles, they argue over who will have the privilege of taking care of us. They resolve that one of them will escort us through customs and the other will be our handler for the remainder of our ten-day trip. Both services, of course, require a modest fee. One of the men, wearing a 1930s Chicago gangster's

PREVIOUS PAGES: *Tarbosaurus,* a member of the tyrannosaur
family, on display at the Mongolian State Museum
in Ulan Bator.
ABOVE: Sergei M. Kurzanov of the Museum of
Paleontology in Moscow with eggs and baby dinosaurs
collected from Russia and Mongolia.
RIGHT: The Flintstones and Rubbles explore Mongolia
OPPOSITE: In our quest for scale comparisons we decided
ducks would be appropriate for a photograph of one
of the largest duckbill skulls ever found. The
skull was found in Mongolia by Russian
paleontologists, and these latter-day dinosaurs
later became their dinner.

hat and what looks like an oversized pajama robe with sleeves twice as long as his arms, wins the argument to be our handler.

I am naturally suspicious of government guides.
—Roy Chapman Andrews,
The New Conquest of Central Asia, 1932

Mongolia has been fought over for centuries. Usually tossed back and forth between Russia and China, it has had many reincarnations since the days of the great and powerful Khans. What changed all this was religion. The Mongolians were shamanists, worshipping the spirits of trees, rocks, and mountains. Then the Chinese advanced Buddhism, and the warlike race behaved like good children and subsequently Mongolia was a country up for grabs.

In 1912, when the Manchu regime in China was overthrown, the Russians seized the opportunity to help Mongolia proclaim her sovereignty. They easily drove out the Chinese. Then in 1914 the Russians were preoccupied with the Bolsheviks, and China, sensing its neighbor's weakness, came to help rescue Mongolia from the Russians. A sort of independence was achieved for a short while, but the Mongols soon tired of the Chinese and enlisted the Russians to drive them out again. In 1921 the Russians assembled a regional all-star brute squad of Tartars, Tibetans, Mongols, Japanese, and Buriats (Mongols educated in Siberia) and extracted the Chinese in a short but bloody war. Mongolia was quickly back in the hands of the Russians, who installed a puppet regime and changed the name of the capital from Urga to Ulan Bator, which means "red city." Because of this last victory most of the buildings in Ulan Bator are built in a bleak, uninspired Communist architectural style that authentically captures all the grimness of Moscow.

When we arrive, Russia is just boiling with internal strife and shedding its satellite holdings. The lack of Russian support and resources has put a stranglehold on Mongolia's already shaky economy, and Mongolians are warming up to the idea of capitalism.

A. Perle, then director of the Mongolian State Museum, opens his arms and his museum for us, and even shuts the dinosaur wing to visitors while we photograph the museum's collections for thirty-five dollars per subject. The museum greatly appreciates the influx of American cash, and we appreciate the financial relief.

One day when we meet our handler, we ask him if there is any way we can get to the Flaming Cliffs, the site where the Andrews expedition first found dinosaur eggs and now perhaps the most famous dinosaur paleontological site in the world.

This is one of the most picturesque spots that I have ever seen. From our tents, we looked down into a vast pink basin, studded with giant buttes like strange beasts, carved from sandstone. One of them we named the "dinosaur," for it resembles a huge brontosaurus sitting on its haunches. There appear to be medieval castles with spires and turrets, brick-red in the evening light, colossal gateways, walls and ramparts. Caverns run deep into the rock and a labyrinth of ravines and gorges studded with fossil bones make a paradise for the paleontologist.

—Roy Chapman Andrews,
describing the Flaming Cliffs in
The New Conquest of Central Asia, 1932

Our handler goes back to his office and returns in the afternoon confirming that there is no traditional way to get to the cliffs. Due to the fuel shortage, all tourist travel has been suspended. However, he says, he does have a friend in the government who might be able to lend us one of the rare working helicopters in the country—for a small contribution. His friend is the prime minister.

THE GIFT OF DEATH

Perle, who was inspired to become a paleontologist by the writings of Roy Chapman Andrews, knows the Flaming Cliffs very well and had agreed to accompany us.

We had asked our handler what to bring the pilots for a gift. He had been bumming cigarettes off John all week, and he had immediately and predictably recommended American cigarettes. We have a whole suitcase full of cigarettes and other gift items for just such

an occasion. At the airport we meet the pilot and copilot, and to establish an immediate bond we offer each a carton. They are extremely grateful and waste no time lighting up. As they hungrily drag on their Marlboros, we inspect the helicopter. I'd rented a lot of helicopters, but none like this one. It's a large Russian-built model, the kind you could transport jeeps in. After every flight a good helicopter pilot will check his craft for signs of a leak and fix it immediately. This helicopter is hemorrhaging various fluids through every possible hose, tank, and connector. Two large auxiliary nine-thousand-liter (2,378 gal) gas tanks strapped on the side are leaking gas onto the tarmac. The tires are threadbare, and wires from the steel belt are protruding from a thin layer of rubber covering. Inside the riding compartment two additional nine-thousand-liter gas tanks drip jet fuel in the narrow aisle onto what looks like a prayer rug. Four stowaways in the back, friends of the pilots, hold handkerchiefs to their noses to stifle the gas fumes. To our horror, the pilots jump aboard still smoking. Through our handler we plead with them to stop. In the few words of English they understand they com-

TOP: The extent of our pilots' safety check before our expedition to the Flaming Cliffs.
ABOVE: Two nine-thousand-liter (2,378 gal) auxiliary tanks leaking jet fuel occupied much of the interior of our helicopter, making it a potential flying Molotov cocktail.

At the Flaming Cliffs (foreground, from left to right): the pilot, our handler, Perle and his editorial assistant, and John Knoebber.

---

municate their response. "No ploblem," they say as cigarette smoke begins to pour out of the cockpit. Our handler, sensing we have given up on the smoking issue, bums a cigarette from the pilot and borrows a light from John.

As the helicopter engine coughs awake and the propeller begins shaking the ancient craft, John says to me, "I never flew in a Molotov cocktail before."

"It may bring a whole new meaning to the Flaming Cliffs," I answer.

"If we get there," John adds, watching our handler flick ashes onto the rug.

Perle pulls a flask of Mongolian vodka from his jacket, takes a swig, and passes it over to us.

> I do not believe in hardships, if they can be avoided, for they lessen effectiveness; they are a great nuisance. . . . Neither do I believe in adventures. Most of them can be eliminated by foresight and organization.
> —Roy Chapman Andrews,
> *The New Conquest of Central Asia,* 1932

Henry Fairfield Osborn, the great curator of the American Museum of Natural History and Roy Chapman Andrews's boss, hypothesized that all mammals and humans originated in Asia and then spread throughout the world. He sent Roy Chapman Andrews, the museum's zoologist, to organize an exploration party.

Andrews wrote: "The main problem was to be a study of geological history of central Asia; to find whether it had been the nursery of many of the dominant groups of animals, including the human race. . . ."

On the morning of April 21, 1922, in five motor vehicles Andrews's exploration party left Kalgan, China, exited the Great Wall, and began their ten-year survey of the Mongolian plateau.

It was the expedition's photographer, J. B. Shackelford, who actually made some of the greatest finds. One day while Andrews was asking some locals about northern caravan trails, Shackelford wandered off and found himself at the edge of a great basin. He walked down the ravine and plucked a skull off a rock, as if it were perched on a pedestal. They camped at the spot, and by dark they had a sizable collection of bones.

Speaking of the Flaming Cliffs, Roy Chapman Andrews wrote:

> It was evident that the formation was Cretaceous and very rich in fossils, but at the time we could do no more than mark it as one of the localities for future work. We could hardly suspect that we should later consider it the most important deposit in Asia, if not the entire world.
> —*The New Conquest of Central Asia,* 1932

The helicopter is filling with cigarette smoke, and we're flying over the suburban yurts of Ulan Bator over the plateau into the great Gobi Desert. We are a few minutes out when the helicopter suddenly lurches. The huge craft rolls on its side, banks ninety degrees, and starts plummeting toward the desert. The helicopter doesn't have seat belts and I have a death grip on the sides of my seat. I look out the window to face my doom. Suddenly the helicopter rights itself and someone grabs my shoulder. It's the copilot. A cigarette dangling from his lip, he points out the porthole window and says, "Volves." I look out the window and there below are two frightened gray wolves scampering across the barren terrain. Our pilot, it seems, is chasing them. He circles again for another pass. I look through the cockpit door. The pilot is laughing demonically, a cigarette stuck between his fingers, one of the female stowaways sitting on his lap. He maneuvers his huge craft through the sky like a semitrailer with

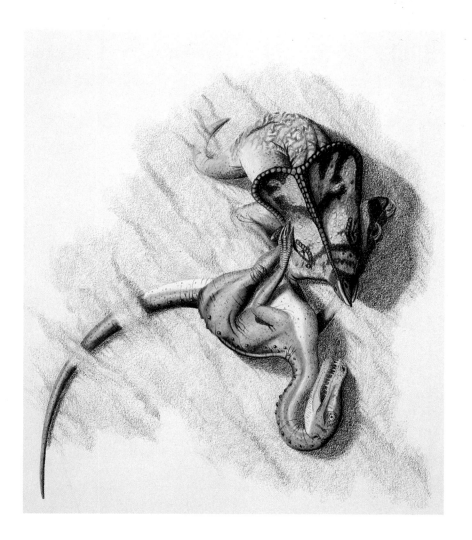

ABOVE: (BEFORE) A juvenile *Velociraptor* attacked a
Protoceratops, which bit down on the predator's right
hand with its beaklike jaws, locking both in a death
pose even after they were covered by a sandstorm.
The *Velociraptor's* hind claw is embedded in the
Protoceratops' belly.
Illustration by Pat Redman
RIGHT: (AFTER) *Velociraptor* and *Protoceratops* as they
were found by Polish/Mongolian
paleontologists in 1971.

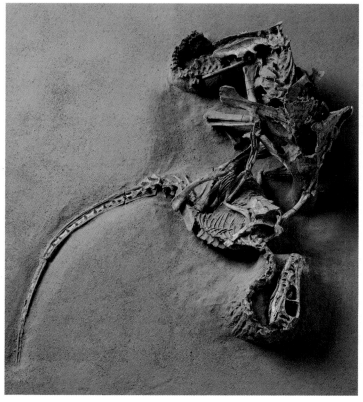

ABOVE: *Oviraptor*—"the egg stealer"
Painting by Shannon Shea
OPPOSITE: *Oviraptor* display at the Mongolian State
Museum.

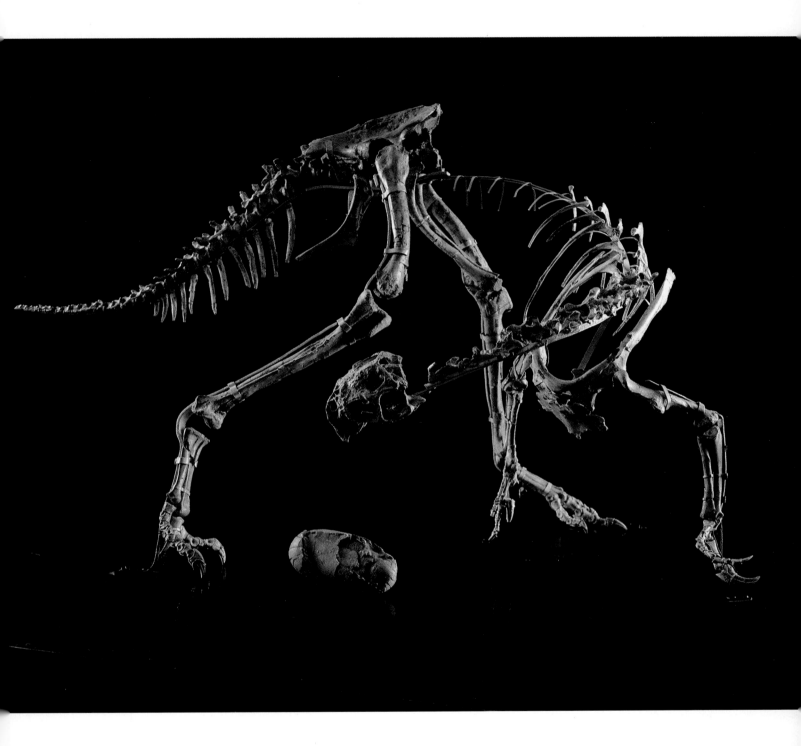

wings. The copilot struggles for balance, nods out the window again, and wiggles his index finger by his eye, a signal for me to take a picture. I then understood that all this cowboy skyriding is solely for my benefit.

We fly low over the fenceless Gobi, passing over wild two-humped camels and herdsmen and their sheep. Andrews's supply camels moved at about two and a half miles (4 k) an hour and averaged about ten miles (16 k) a day. With a refueling stop, we make it to the Flaming Cliffs in three and a half hours. We land in the desert on rocks the size of grapefruits. Miraculously, the tires hold. It was our intention to make a pilgrimage to the Flaming Cliffs, but we aren't prepared to collect specimens. We arrive when the light is still high in the sky and poor for photography. However, it's great light for prospecting.

About ten minutes into our hike, at the base of the cliffs, Perle notices a claw on the ground. "Velociraptor," he says, and walks on. On August 12, 1971, with two colleagues from the Polish/Mongolian expedition, Perle helped excavate the famous specimens of *Velociraptor* and *Protoceratops* that died locked in mortal combat and were then buried by a sandstorm. After finding those specimens, I imagine, finding a single *Velociraptor* must be pedestrian to him. As Perle strides on, I brush some dirt away from the claw with my hand. The *Velociraptor* hand is articulated, each bone in

continuation with the next, and it looks like it's reaching to pull itself out.

But I have to abort the excavation because I don't want to lose track of the fast-moving Perle, who is already over the next rise. As I follow him up a hill to the top of the cliffs, I stumble onto a dinosaur nest. The eggshell has eroded, leaving only the forms. I don't want to bother Perle with such a minor find, but on the way back down, since we are passing the spot anyway, I show them to him. He stares in disbelief. "I congratulate you, Louie. Excellent find," he says. We harvest the eggs for the museum by simply plucking them off an outcrop and putting them in our jacket pockets.

In two hours we had found a nest of dinosaur eggs and a *Velociraptor.* Not bad for an afternoon's work.

Shortly after we got back from Asia, I started writing this book in Antigua, far away from anything to do with civilization or dinosaurs. One afternoon I decided to take a break from writing and watch the sailboat races from the top of the cliffs near my house. A lone photographer was there taking pictures. His name was Alexis Andrews, an Englishman of about my age living in Antigua. We had met before socially. He asked what I was up to, and I told him that I was writing a book about dinosaurs. "Really!" he said. "Have you heard of my grandfather—Roy Chapman Andrews?"

OPPOSITE: One of several egg forms discovered by the author at the Flaming Cliffs of Mongolia.

WHERE THE DUCKBILLS ROAM

JACK HORNER'S HUNT FOR BABY DINOSAURS

A couple of hundred yards from the teepees, past the flapping American flag over the kitchen shack and beyond the latrines, Jack Horner, who received a "genius grant" from the MacArthur Foundation in 1986 for his work on dinosaur nesting sites, takes me over a hill and down a ravine to look for dinosaur eggs. In the distance, the Rockies ripple through heat waves rising off the badlands. Jack lays belly down on the baked ground, his face six inches (15 cm) from the dirt, and begins staring intently for several minutes, barely moving, like someone who has lost a contact lens. Eventually I interrupt to ask him what he sees. Jack has a severe case of dyslexia and takes a few seconds to respond while he processes the information. He picks up a black, fingernail-sized sliver of something and hands it up to me.

"Dinosaur eggshell," he says to my astonishment. The one-millimeter-thick (0.039 in) piece of fossilized shell is slightly curved and has a pebbly surface just like a modern chicken egg's. Under pressure it snaps apart with the brittleness of a modern eggshell.

"We give it away to visitors as a souvenir," he says as he sits up and wipes the sweat from under the brim

OPPOSITE: Jack Horner at his Choteau, Montana, teepee encampment.
Photo by Louie Psihoyos © National Geographic Society
ABOVE: An *Orodromeus* egg with the bones of an embryo inside.

of his hat with the back of his hand. "Dinosaur eggs are probably everywhere. They are probably in every formation. I'm sure they are; we keep finding them. But you can't walk around in a standing position and just see eggshell fragments on the ground. It's too hard to see them. You have to get down on your hands and knees and look."

I take his advice. After several minutes of crawling on my knees over sharp caliche, I accumulate a small handful of dinosaur eggshell, first broken, Jack tells me, some 80 million years ago by hatchling baby dinosaurs on their first day of life.

In 1978 Jack Horner was a preparator for Princeton University on a working summer vacation looking for baby dinosaurs in his home state of Montana. With him was his buddy Bob Makela, a Montana schoolteacher. Like many other paleontologists who came before them to search for baby dinosaurs, they were without success. A colleague, Bill Clemens of the University of California at Berkeley, whom they had helped to find some mammal remains, mentioned that a rock shop in another part of the state near Bynum, Montana, needed some dinosaur specimens identified.

They did their good deed, and as they left the proprietor asked, "By the way, I have some little bones. Can you tell me what they are?"

ABOVE: Jack Horner looking for eggs at
Egg Mountain in Montana.
BELOW: Bozeman (A), Egg Mountain (near Choteau) (B),
and Bynum, Montana (C).

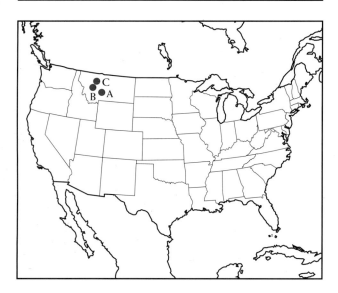

Jack shakes his head in amusement, remembering his extraordinary good fortune. "She had a whole coffee can full of baby dinosaur bones," he says. "There were at least four babies represented. The owners didn't know they were baby bones! They had planned on sticking them to a little piece of cardboard and selling them for a quarter apiece. I told her they were important and she let us have them."

The rock shop owners, the Brandvolds, showed Jack and Bob the site, later dubbed Egg Mountain, where they got the little bones. It was about fifteen miles (24 k) west of Choteau, Montana, on a secluded ranch nestled between two nuclear missile silo sites near the base of the Rocky Mountains, where Jack was now showing me the eggshell.

Jack describes the first eggs he and Bob Makela found. "We dug down and found them in a bowl-shaped depression, so we assumed it was a nest. It was about six feet [183 cm] in diameter and about two and a half feet [76 cm] deep. There were eggshell fragments and plant debris in there, too, with some terrestrial snails."

After Bob and Jack dug out the nest, they went to Jack's hometown, Shelby, fifty miles (80 k) away, and were sitting in a bar talking excitedly about their find when someone next to them overheard their conversation and tipped off the local press. A wave of publicity began that still continues today. Jack explains his simple equation for success. "More publicity brought more funding, and more funding brought more discoveries."

A single dinosaur nest was a spectacular find, but in subsequent field seasons in '79 and '80 Jack and Bob Makela started finding more nests on the same geological horizon, eight in all, each nest about twenty feet (6 m) from the other. Their proximity to one another led Jack to popularize a concept then foreign to our image of dinosaurs—that they evolved the instincts to nurture their young.

"Since the average *Maiasaura* is about twenty feet [6 m] long," Jack explains, "my assumption was that this was a tightly packed nesting colony. We were only able to follow it over a relatively small area, but it covered about ten thousand square meters [2.5 acres]. Some of the other nesting grounds we found are huge things. The *Hypacrosaurus* [another kind of duckbill] nesting ground is about a mile and a half [2.4 k] long and a half a mile [0.8 k] wide."

He named the new duckbill dinosaurs *Maiasaura*, which means "good mother lizard." "It was the first dinosaur name to be given a female gender," Jack tells me.

By slicing the baby bones into paper-thin cross sections and studying them under a powerful microscope, Jack determines that baby *Maiasaura* were in the nest for about a month. He explains, "The ends of their bones are made up primarily of calcified cartilage, rather than ossified bone. As long bones grow, the end of the bone continues to grow and the inner

ABOVE: Artist Matt Smith's reconstruction
of a *Maiasaura* hatchling.
OVERLEAF: *Maiasaura* nesting ground
Painting by Pat Redman

part of that cartilage calcifies and becomes regular bone. If a bone is growing real fast and isn't needed to walk around on, then the calcified cartilage can be really huge. So when you find a bone where there is just a little bit of ossified bone, you can tell how old it is. That's the same situation you see in altricial birds that don't go anywhere but the nest: they can't walk. So if baby maiasaurs were in the nest for four weeks and couldn't walk, the assumption is that they, too, had to be cared for."

Jack suggests that adult dinosaurs may have been conned into becoming good parents by one of Nature's tactics he calls the Cute Factor—when a baby is so irresistibly cute it triggers an innate nurturing instinct. Even so, maiasaurs, he explains, probably weren't cute for long.

"Dinosaurs started out like birds," he says, "with really rapid growth which allowed them to get big quick. That's probably why baby dinosaur finds are rare, because they didn't stay babies very long. Their growth spurt is very fast when they're young, and then they shut their metabolism down when they're big. A baby *Maiasaura* hatched out of its egg was eighteen inches [46 cm] long, and at the end of four weeks they were about three and a half feet [107 cm] long. They grew to nine feet [274 cm] in the first year and about twenty feet [610 cm] in about four and a half years. I suspect that baby dinosaurs probably were down-covered, but when you get twenty-five feet long, the last thing you need is insulation. Twenty-five feet [762 cm] wasn't the maximum, though. They kept growing throughout their whole life, but at a point, bone deposition slows way down—probably when they reached sexual maturity. Dinosaur growth rates are magnitudes higher than anything, as far as we know, that can be achieved by any cold-blooded animal."

Concrete evidence for the massive herds of Maiasaurs that Jack envisioned came soon after the discovery of the first nesting sites. And again the evidence came disguised as dumb luck.

When we return to camp, I notice for the first time some huge blue tarps at the edge of Jack's teepee village. Jack tells me the tarps cover a bone bed called Camposaur. Camposaur is a few seconds' commute from any teepee and right next to the mess tent. Even the most dedicated paleontologists usually like to distance themselves physically from their work, at least a little. Jack explains, "Camposaur was found in 1981. Our crew kept getting bigger and bigger, so we kept looking for places to put tents. We had this low hill right next to camp, so we told people they could camp there. One guy tried to put his tent there, but he couldn't get his stakes in the ground. We thought that was sort of unusual because it should have been relatively soft ground. We brushed away some of the dirt, and there were some dinosaur bones in there, about four or five. So he had to move his tent, and we asked him to dig up the bones. He found about twenty or thirty more bones and, well . . . we are still working on it. It turns out it's the biggest bone bed in the world, and we had been camping on it for years without realizing it. It's a little embarrassing. It's about a half mile [0.8 k] wide, and conservatively it's three miles [4.8 k] long."

He explains that several of the *Maiasaura* bone beds in the area he had been working on all turned out to be parts of one humongous bone bed. "There's ten thousand individuals in there, but that's a conservative estimate," he tells me. "People are still trying to figure what happened, but it appears that it was a *Maiasaura* herd that died in a catastrophic volcanic event—they were too close to a volcano that was erupting. You can tell how hot it was by the size of the ash. Geologists have estimated that the ash was somewhere between five hundred and six hundred degrees Fahrenheit [260–316°C]. If we use Mount St. Helens as a comparison, the animals that died, like the elk, the meat on their body was literally cooked off the skeleton. So you end up with a lot of bones very quickly."

All the nesting sites he found, although they were within a few miles of the Rockies, were once part of a coastal plain when the Gulf of Mexico came up into Canada. Paleontologists call this ancient ocean the Colorado Sea. The sea, only a couple of hundred feet (61 m) deep, came and went several times during the Mesozoic, forcing the lowland coastal dinosaurs, when it rose, into an upland habitat. "It appears that they were habitat-restricted," he says. "The Rockies were to the west and the Colorado Sea to the east, so it appears that they probably would have migrated north seasonally."

And when in the north, it appears they laid their eggs.

"In 1979 we were camped out on the Teton River and would drive about twelve miles [19 k] to the Willow Creek anticline. We'd walk out to a site we called Sacred Slump, which was a maiasaur skeleton we were digging up. One day we saw some people coming across the prairie sticking little red flags in the ground. And so we asked them what the hell they were doing out in our place where we were digging, and they said they were laying a seismic line; they were going to bring some big thumper trucks through and blast to see if there was oil under the ground. I thought it would be a good idea to send some people down the line and make sure that there weren't any good dinosaurs that they were going to blow up. One of the students in our crew, Fran Tannenbaum, was tired, so I sent her on the line that headed for our truck. When she followed it, she found one of the red flags sitting next to an egg. She started yelling and screaming. We ran over there, looked at her egg, then started looking around the hill, and we found about a dozen eggs that

ABOVE: Jack's teepee encampment at sunset
BELOW: At Camposaur, Jack discusses fossils discovered by an up-and-coming paleontologist.

day. The first one she found was *Troodon,* and most of the rest of them were *Orodromeus.*"

With a string of successes behind him, Jack eventually left Princeton and migrated back west to Montana, where he became the curator of paleontology for the Museum of the Rockies. Under his direction, the museum excavated thousands of new specimens. His museum, however, is unlike other dinosaur museums in that there are very few bones on display.

"Every other museum has skeletons on display," Jack tells me. "I just don't like them. Skeletons do not portray life, they portray death. I know that skeletons are really neat to look at. The kids love to see how big the skeletons were, but in our hall what I am trying to do is portray dinosaurs as living animals, not as just the ideal of extinction."

Instead Jack shows dinosaur sculptures and some realistic Japanese robotic dinosaurs. "I want people to see dinosaurs as they would have been when they were alive," he explains. "Also it's hard to interpret the life of a dinosaur by the skeleton. To give you a good example, I once took a mounted skeleton of a horse to a horse show, and the people there thought it was a dinosaur. And these are people who know horses. Most people cannot envision what a fleshed-out animal would look like from looking at a skeleton. Consider a rhinoceros. It doesn't have a horn on the skeleton. The horn is just a big wad of hair. Dinosaurs may have had a lot more frilly parts, but we don't know that from the skeleton either. Basically, from the skeleton, you can only make a conservative animal."

Later that evening around a campfire under a ceiling of stars, Jack, with Twainian wisdom, is talking about dinosaurs to a group of about forty devoted youngsters, housewives, and other would-be paleontologists.

A student asks Jack, "Did any of the dinosaurs hibernate?"

"No. It would take a lot of big holes," Jack answers.

An older student asks, "When they were exercising or moving fast, how did they prevent overheating?"

Jack thinks about this and says, "Damned if I know. If I were them, I wouldn't move fast."

With a hurt, almost accusatory voice, a teenage student asks Jack, "I hear you're collecting mammals now?"

Jack responds, "The theme of the Museum of the Rockies is 'One place through all of time.' We are making a Paleozoic Hall and a Cenozoic Hall. We need some mammals for the Cenozoic Hall. Obviously right now the vertebrate paleontology collection is slightly biased toward dinosaurs. We have thirteen thousand dinosaurs and about a dozen mammals and one invertebrate. So in order to make a Cenozoic Hall, we are kind of forced to go out and collect the overburden."

Another student asks, "Why is Montana such a wonderful fossil collecting ground?"

"Well, because dinosaurs only died in beautiful places," Jack says. "And because two-thirds of the state's surface is the right age of rock for looking for dinosaurs."

Another student asks, "Dinosaurs weren't very smart, were they?"

Jack is quiet for a moment and finally says, "Intelligence is an interesting thing, because people are biased. We try to measure an ability of an animal by its intelligence. But intelligence is the only thing we humans have going for us. We can't run fast. We can't see or hear all that well. The only thing we can do is think. So we're real concerned about how well other animals can think. A bald eagle doesn't have to think much. But it certainly needs good eyes. If you look at the brain capacity of a dinosaur, and compare it to animals alive today, you'll notice that it has a relatively small brain but it has a very large olfactory lobe, which means it could smell anything. It also could hear very well and some of them could see very well. So they are all very well adapted to their environment. And since there wasn't anything alive at that time that was walking around thinking about making a civilization and cutting down rain forest, it didn't have to worry about out-thinking its neighbor. All it had to do was find food and do its best to stay away from its enemies.

"It's unclear whether intelligence has done us any good. It's still unknown if intelligence would have done a dinosaur any good. We have to be able to figure out whether intelligence is a good thing to have before we start judging other animals.

"Did you see the missile silos on your way up here? If those missiles went up and others came down on us, there would be two piles of bones here—the maiasaurs' and ours. Then which of the two species would you say was the more intelligent?"

THE EGG MAN

Karl Hirsch is considered the world's best authority on dinosaur eggshell. He described the first fossil turtle eggshell, the first fossil crocodile eggshell, and has written more papers on the subject than anyone. I talked to him at his home in Denver.

"You could also say I was the *worst* authority because I'm the only one," Karl says. "I've worked almost twenty years on eggshells. But I did it as a hobby. I wanted to find out more, and the more I learned, the more I got hooked.

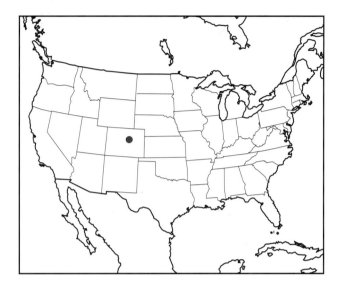

RIGHT: Denver, Colorado
BELOW: Karl Hirsch, the world's leading expert on eggshell of all types, with an artist's reconstruction of a sectioned *Orodromeus* egg with embryo.

"I worked in maintenance in a building where we processed plutonium at Rocky Flats. We kept the place going. We repaired machinery and I worked nights, and during the day I was a rock hound with my wife. One day we went collecting in the badlands. There was a storm, and we had to leave the jeep there and hike in. A fossil egg was laying there, clean and washed like somebody put it there for me. I went to various museums, without success, to have it identified, and then I wrote to Alexander Wetmore, curator [now deceased] of the Smithsonian Museum of Natural History, and he gave me some of the references on eggshell literature. And there wasn't much. He said that if I wanted to know more I would have to do the rest myself.

"My wife and I, we weren't scientists, but we tried to find out what we could. I've written some twenty-seven or twenty-eight papers on the subject of eggs since. I have studied reptile eggs, bird eggs, turtle eggs. Now people bring me eggs from all over the world.

"Birds, they have an assembly line in the uterus. That egg yolk goes in the uterus and they put a membrane around it, and then in another place the nucleation points come out, and then in another place it's coated with an eggshell. So the chicken has to be first, in this case. And then they lay the egg, and then the other one is processed.

"I explain to kids that an eggshell is like a house where baby dinosaurs are protected from the outside world. The shell is like a screen so no insects come in, but the pores on the shell let good air in the window, and the bad air, CO_2, goes out. Now to identify who lived in that house—identify that eggshell—it's as if you go over to Europe, you almost can say which country you are in by the way they build the houses. But the only proof, hundred percent proof of who laid that egg, is to find the embryo. Right now there are only three one hundred percent identifiable, confirmed embryonic remains within an egg—a *Maiasaura* and *Orodromeus*, both from Montana, and a *Hypacrosaurus* from Devil's Coulée, Alberta. Those are the only ones I would put my hands in the fire for."

I ask Karl if he thinks that one day scientists might be able to clone dinosaurs. He pauses and strokes his chin thoughtfully. "I don't think they should clone dinosaurs even if they could," he says. "They should try to clone a good politician—that would be even more rare."

"When I started working for Jack Horner, I was his histological technician, doing thin-sections of bones," says Karen Chin, the world's expert on dino dung, as we stood around a campfire in the Montana Rockies. "Before this I had been a naturalist for fifteen summers in the Park Service. I was a naturalist in Kings Canyon National Park, in Yellowstone National, and for ten years in Glacier National Park. And when you're a naturalist you look at everything, including scat.

"I knew scat can tell you a lot. It can tell you a little bit about the diet, it can tell you about the populations of animals, and, within reason, who made it. Sometimes it's difficult to tell black bear from grizzly scat, but you can narrow it down to bear with no problem. The diameter gives you an idea of how large the animal was, or you can get a really good idea of what its diet was. Some bear scats you see are totally red, loaded with buffalo berries. Or other times they're loaded with grass—they eat a lot of grass or cow parsnip. And so you can see exactly what these bears have been eating. It's a real important way of interpreting what kinds of animals that you can find in an area.

"A coprolite is fossilized feces. So when Jack told me he had some dinosaur coprolites, I got . . . well, I was real interested and asked him if I could make a thin-section, just like I was doing with the bones. And I did, and it was really neat because it was a way to link the plant community, which I've always been interested in, with the dinosaurs themselves. Nobody has ever done an exhaustive study on coprolites. Most geologists wouldn't want to put their name on it.

"Well, you know it's dinosaur poop by the shape. And that's still the best criterion. It looks like poop. That's it. It looks like poop. For the carnivores it's mineral composition. It has a calcium phosphate com-

OPPOSITE: Karen Chin, the world's expert on fossilized dinosaur dung (coprolites), with her collection of suspected droppings.

ABOVE: There are an estimated 10 to 40 million species of beetles, representing about 10 percent of all known life-forms. Lineage of the beetle can be traced further back than the dinosaurs. Today there are about ten to thirty thousand species of dung beetles, some of which have been eating dung since the Mesozoic. The *Heliocopris dominus*, a modern scarab dung beetle from Thailand, shown above, adapted its specialized headgear for the same reasons as the *Triceratops*—to attract mates and aggressively protect its food resource.

Illustration by Pat Redman

LEFT: From a book of insect illustrations, circa 1770, by Dru Drury.

position, and sometimes it has bone in it. A lot of carnivores now crush bone to get at the bone marrow. And there's no herbivore then that could have been making anything that big that could have been eating such a fibrous diet. The mammals were all tiny, no bigger than rat size, if that big."

Chin explains that for coprolites to be preserved, they need to be permineralized. Permineralization is the process by which fossilization occurs. Through many years groundwater percolates through a bone, or in this case dung, and the minerals left behind, often silicates, fill the pores and preserve the fossil with a rocklike structure.

"It's easier to permineralize bone," she says, "because you have a hard substance to start with, and it will last a long enough time to allow groundwater to seep in and fill up the little holes. With soft tissue, like feces, you don't have as much time to let the stuff sit around, because it is a source of food for other animals and microorganisms.

"There is a dung beetle expert, Bruce Gill, from Toronto, who told me that dung is a real exciting food source for a lot of animals. Elephants defecate and almost immediately there are beetles out there that are cruising, flying around just waiting for big animals to defecate. And they have such incredible scent receptors that within a few minutes of an elephant or big herbivore defecating in Africa, it's covered with dung-eating flies and beetles. There's enormous competition.

"Anyway, it's an incredible resource. I have found very, very few bona fide herbivore coprolites in the fossil record from the Mesozoic, and you figure there's got to be a couple of reasons; one, they preserve very poorly, which is a valid possibility, or number two, there was a lot of coprophagy [feces-eating] going on. In Africa the stuff disappears like that. So anyway, when we look at the traces and this dubiocoprolite [suspected doo doo], there are some burrows and some other evidence in there that, Bruce says, is indisputably dung beetle. There's no doubt. And you know what this means?"

I admit I don't.

"This is the biggest verifiable herbivorous dinosaur coprolite found. Am I yelling? I'm getting excited. As much as I can tell from looking at the literature, it's the biggest one known. Confirmed. This may well be the first documented dinosaur–insect interaction, which tells us something about the food web. It tells us there is a dinosaur, probably *Maiasaura*, feeding on conifer twigs, and there are beetles feeding on this dinosaur's dung. We now have the beginning of a Mesozoic food web!

"How the dung beetle survived the Cretaceous/Tertiary extinction without dinosaur dung, I don't know. Perhaps the dung beetles were moving to the pile of a different dunner."

SONG OF THE CRESTED DUCKBILLS

"I'm not sure you want to say David Weishampel is best known for his honker," Dave Weishampel says. "Why are you laughing?"

I don't want to tell him he just wrote my lead.

It's before opening hours at the Royal Ontario Museum in Toronto. I'm at the hadrosaur collection with Weishampel, a vertebrate paleontologist from Johns Hopkins University, as he puts together his honker, a device he created to duplicate the sound of a *Parasaurolophus*, a strange crested duckbill dinosaur.

"What does it sound like?" I ask, trying not to laugh.

He thinks for a moment and says, "Like long, sustained belches, the kind that would make any thirteen-year-old boy in a locker room proud. I did the mathematics, did the acoustical physics stuff, where you put numbers in and get numbers out, which came out pretty reasonable. In conjunction with a BBC television program, I constructed this model out of plastic tubing so I can actually replicate the sound of a *Parasaurolophus*. Unfortunately, this is what I'm going to be best known for."

He continues. "Principally the crests of these hadrosaurs are functioning as resonators in vocalization, much like trombones. Any kind of musical instrument is a resonator, much like nasal cavities, and various sinuses and mouths are resonators. These hollow-crested hadrosaurs, known as lambeosaurs, on the face of it had done a really peculiar thing with their nasal cavities. Whereas everybody's conventional nasal cavity, yours and mine, and flat-headed hadrosaurs', are lodged in the

face directly in front of the eyes, lambeosaurs have taken that nasal cavity, plucked it out of their faces and stuck it on top of their heads and made it variously shaped, and provided it with greater length. What it does is enhance the lower frequency, the low notes of any kinds of sounds that they may have made."

"For mating purposes?"

"Presumably mating, warning, communication between the little guys, hatchlings, teenagers, and Mom and Dad, comments about the ripeness of fruits and leaves and stuff like that. The general basic kinds of communication that are represented in virtually all animals, I can assure you, were the kind of things that hadrosaurs did as well. These guys didn't have ears that stick out like mammals do, like we do; they had an eardrum lodged onto the side of their heads like birds, crocodilians, and lizards. It's reasonably large and deep, and bumping against the eardrum is a long, spindly bone called the stapes, which is very conducive to low sound waves. In fact it would amplify them quite a bit. The inner ear itself is much like a crocodilian's that's associated with hearing low frequencies. For instance, crocodilians during the breeding season regularly produce sounds and hear sounds that are roughly the frequency of your lawnmower, and react to them territorially. But they also hear reasonably high frequencies, because the full-term embryos inside eggs actually make cheeping sounds which triggers Mom to pull the vegetation off the nest, pick up individual eggs, crack them gingerly with her teeth, and release the hatchling from inside."

Having finished putting his fabrication of a duckbill's nasal resonator together, he announces, "This is what the *Parasaurolophus walkeri* type specimen would sound like."

His cheeks puff up like Dizzy Gillespie, and suddenly a long, sustained belching sound reverberates throughout the halls of the museum like a whole herd of duckbills with gas pains. His Cretaceous symphony summons the territorial response of a security guard a few minutes later. By way of explanation, Professor Weishampel performs an encore.

The security guard looks hard at Professor Weishampel and then turns to me, perhaps assuming I am the sane one, and asks, "Are you guys going to be out of here by the time the public arrives?"

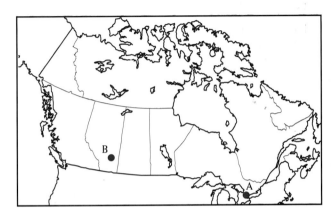

ABOVE: Toronto (A), and Dinosaur Provincial Park, Alberta (B), where the type specimen of *Parasaurolophus* was discovered.
BELOW: A chilly Late Cretaceous morning finds *Parasaurolophus walkeri* wandering through a conifer forest. A sudden forceful exhalation of air through modified nasal passages in the animal's hollow crest produces a long, low, trumpeting call.
Illustration by Michael W. Skrepnick
OPPOSITE: David Weishampel with his honker, an instrument that simulates the sound of a *Parasaurolophus*, a duckbill dinosaur with a customized nasal passage.

THE JOY OF *REX*

PETE LARSON'S RENDEZVOUS WITH NATURE'S TERMINATOR— A *TYRANNOSAURUS* NAMED "SUE"

Many, if not most, of the great dinosaur finds have been discovered by amateur collectors. Witness all but one of the fifteen known *Tyrannosaurus rex*. Strangely, until recently *T. rex* was one of the least understood dinosaurs, because the specimens have been relatively incomplete. Then "Sue" was found. Sue, the biggest and best *Tyrannosaurus rex* ever discovered, was nicknamed after Susan Hendrickson, an amateur paleontologist and adventurer who found her near the town of Faith, South Dakota, while collecting with the Black Hills Institute, a privately owned company run by fossil collectors Pete and Neal Larson and their buddy Bob Farrar. The unique quality of Sue's preservation was leading scientists to reveal incredible stories of her life and tyrannosaur behavior—right up until the F.B.I. came and seized her on an early morning raid of the institute.

SUE FINDS "SUE" ON SIOUX LAND

August 12, 1990, had started out more miserably than most for Pete Larson. When he woke up he noticed his truck, the only vehicle at the remote site, was starting to get a flat tire. The spare, however, was flatter than the flat, and when he tried to pump up the tire, he found that the tire pump was broken. He decided to drive into town with the crew in the hobbled vehicle to get the tires fixed. Sue Hendrickson was a volunteer who, like many other collectors, wanted the opportu-

OPPOSITE: "Sue," the largest and most complete tyrannosaur ever found, with Pete (left) and his brother, Neal Larson. Sue was named after her discoverer, Sue Hendrickson, who found her three miles from the Larsons' Ruth Mason Quarry on August 12, 1990.
ABOVE: Site of Sue excavation, near Faith, South Dakota.
Photo by Pete Larson

nity to work with talented scientists and the chance of finding something great. The week before they had secured permission to look for fossils on some nearby ranch land owned by a Lakota Sioux Indian, and she decided to stay behind and prospect. Susan was three miles (5 k) from the dinosaur quarry when she came across some fragments of bone spilling from a fifty-foot (15 m) cliff.

"When Susan found Sue," Pete explains, "she first saw fragments of bone lying on the ground. She looked up about seven feet [2 m] high on a cliff and saw a cross section of bones about eight feet [2.4 m] in length weathering out of the cliff face. Luckily it was a cliff face, so it weathered fairly slowly. But we still have more than ninety percent of Sue's skeleton."

After Sue's discovery, the scientific world was buzzing with rumors that the Black Hills Institute, which sells fossils as teaching specimens to schools and museums around the world, was going to sell their tyrannosaur for sixty-five million dollars to the Japanese. John and I decided to pay a visit ourselves.

The institute is in Hill City, South Dakota, a Norman Rockwell–like town just a few miles down the road from Mount Rushmore. Despite its name, Hill City is in a valley. The principal enterprises of the area are tourism and logging, both hit hard by the recession. An old steam train at the edge of town carries tourists up the valley, and when its lonesome whistle wails and reverberates between the turn-of-the-century storefronts of Main Street, it's easy to feel you're in a different century. The biggest building on Main Street is the former American Legion Hall, now occupied by the Black Hills Institute.

Pete Larson, who is as passionate about fossils as anyone I have met, shows me around and reveals to me their private fossil collection, specimens he says aren't for sale at any price. They, along with Sue, he explains, are going to be put in a museum they plan to build right there in Hill City, a museum that will revitalize the town.

"I've been working my entire life with the dream of building a museum," he says. "My brother Neal, Bob Farrar, and the whole staff here have been working with that dream. You saw a picture of our first museum when I was eight years old. Sue is the cornerstone of that museum. From the moment we found her we said Sue wasn't for sale. From the first press release we said, 'Sue is to be the anchor of the Black Hills Museum of Natural History.' This dinosaur will make that museum happen.

"We will be rich beyond our wildest expectations in the museum that we will have. I'm not talking about money-rich. I'm talking about rich in fulfilling our dreams. Money, after all, is only fun tickets.

It buys you what you want. Why in the hell would I sell a *Tyrannosaurus rex* when that's the thing I have most wanted in my whole life?"

When the Larsons were growing up on their father's ranch on the Rosebud Reservation in South Dakota, Pete tells me, they didn't play cowboys and Indians—they played curator.

"I started collecting fossils when I was four years old," Pete Larson says. "Not dinosaur bones but mammals. I was about ten when I discovered my first major fossil—a tooth of a gamphothere, a primitive elephant. I found the tooth in a gravel pit near the Little White River."

Unfortunately, the elephant had been bladed over by a bulldozer. Pete took home all the pieces he could find, and working nights after school at the kitchen table, reassembled the large tooth and restored the missing parts. He sent a letter and a picture of the

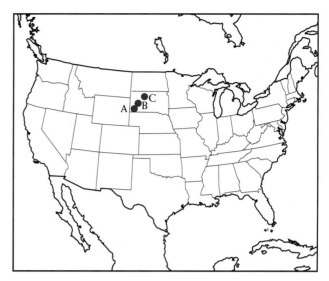

ABOVE: Hill City (A), Rapid City (B), and Faith (C), South Dakota, the site where Sue was discovered.
OPPOSITE: The Larson siblings and their first museum at their childhood home on Rosebud Reservation. Pete, age eight; Neal, five; Mark (big guy), eleven; and sister Jill, age two.
Black Hills Institute

specimen to Robert Wilson, then director of the Museum at the School of Mines in Rapid City, who identified the specimen and encouraged the young Larson to pursue his efforts. Larson entered his elephant and a number of other fossils in the State Science Fair and won first prize for his paleontological preparation.

Pete continued his collecting, received some scholarships, and eventually entered the School of Mines, not because he wanted to be part of the mining industry but because he loved fossils. Fossils are studied by geologists, not out of a love of nature but as an indicator of oil and gas potential. Larson didn't like the idea of working for the big energy companies and was always looking for a way to stay involved with fossils.

While at the School of Mines, Larson met Bill Roberts, the curator of mineralogy, who wrote the *Encyclopedia of Minerals*. Pete Larson and his pal Jim Honert would go into the Black Hills with Roberts and help him in the discovery of a number of rare and new minerals. Roberts, who would invite the students over to his house and quiz them about his vast mineral collection, told them it had been his dream to start an earth science supply house that would provide teaching specimens for schools and universities.

The Lizard Queen's Final Day,
by Michael W. Skrepnick.
While resting in the warm late-afternoon sun, in what we would now call South Dakota, an aging, vulnerable, and battle-weary *T. rex* that paleontologists will name Sue alerts her family members to the appearance of intruders.

Sue has survived a number of injuries in her long career as the region's top predator, including fractured bones, torn muscles, infected bites, and a broken tail. These injuries all bear testimony to previous brutal encounters, many with members of her own kind. She slowly rises to her feet, bellowing in discomfort; her younger mate is

already moving toward the edge of a nearby stream, ready to challenge the potential transgressors. Seemingly oblivious to the impending danger, one of Sue's offspring concentrates on launching a mock attack (a test of hunting prowess) on an unsuspecting eubaenid turtle. Several minutes later, after a vicious assault by a large

group of marauding rivals, Sue and her family are left dying at the water's edge. Eventually her carcass and parts of the others are buried under sandy sediments and carried downstream by the strong, unrelenting currents.

"We thought this was a pretty good dream," Pete says. "And we thought this might be a way in which we could take our interests in mineralogy, geology, and paleontology and be able to do that for a living."

With another pal, Larson started up a business in 1974 under the name Black Hills Minerals. They began selling fossils as a way to pursue their passion and pay their way through college. But, Pete tells me, his master's adviser thought the practice of selling fossils was immoral. Pete had some scholarships to get through college, but still he had to work to pay his way through school, and since he was just getting their business off the ground, it took him longer than most students to complete his course work.

"I finished my courses, and when it came time to do my master's thesis," Pete says, "I talked to my adviser and he said, 'Well, you know, in the catalogue here we have a little stipulation that says that if you take more than five years to finish your thesis, you *may* be required to take some of your courses over. *You* will be required to take *all* of your courses over.' I went to the head of the Geology Department, and he said, 'Well, he's your adviser, what he says you have to do.' I went to the dean of students and he said, 'Well, he's your adviser.' So I just left."

Despite not getting a master's degree, Pete put his energy into his business. His partner eventually sold his share of the business to Pete and Neal. Then Pete, Neal, and their buddy Bob Farrar joined forces to found the Black Hills Institute of Geological Research.

Today there is no institution anywhere in the world that has fossils that rival the beauty of those at the Black Hills Institute. They are pioneers in the complicated art of preparation, the removal of rock surrounding a fossil. Good preparation can make or break a good fossil, and their institute spends more time and energy on the final stages of preparation than any organization, public or private. Fossil preparation is a skill that blends science, art, high technology, and superhuman amounts of patience.

PRACTICING PALEONTOLOGICAL RELIGION WITHOUT A LICENSE

Even though Pete, Neal, and Bob sell fossils commercially, they are also accomplished scientists and were inviting other scientists to study Sue. They contacted

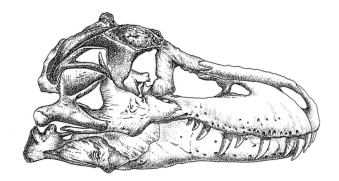

ABOVE: Sue's skull
Illustration by Dorothy Sigler Norton
BELOW: Children of Hill City, South Dakota, protesting the removal of Sue outside Black Hills Institute. Sue was to be the centerpiece of the Hill City Museum.
OPPOSITE: Terry Wentz, Sue's preparator, told us, "After sixty-five million years you can still cut yourself on a *T. rex* tooth." *T. rex* had serrated edges on the fore and aft edges of its teeth, and like sharks they constantly rejuvenated teeth throughout their lives.

Andrew Leitch, a Canadian-based paleontologist who is using medical CAT scanners to nondestructively probe the internal workings of dinosaur skulls. Previously if scientists wanted to probe a skull, they did it the old-fashioned way and broke it open. By placing a small skull of a dinosaur called *Nanotyrannus* in a scanner, Leitch and paleontologist Bob Bakker had recently found turbinals, sinus bones, that extended the amount of receptors of the olfactory bulb and indicated that the beast had a keen sense of smell. Using a coat hanger as a crude probe, Bakker had later found that the massive skull of a *T. rex*, like that of modern birds, was actually filled with a labyrinth of tunnels that

helped lighten it and may have provided, Bakker thinks, an air-conditioning system for its brain. By placing Sue's marvelously preserved skull in a CAT scanner, the Black Hills folks hoped Sue would reveal even more secrets. The problem was that the five-foot-long [152 cm] skull was too big for traditional CAT scanners.

The Black Hills Institute then found an odd alliance with the space industry. They planned to take Sue down to the Marshall Space Flight Center in Huntsville, Alabama, and pass her skull through a powerful industrial CAT scanner used by NASA to test for cracks in space shuttle engines. I was packing my bags for the trip when I decided to call Pete Larson to see if we were still on schedule.

Terry Wentz, Sue's chief preparator at the Black Hills Institute, answered the phone. I asked him if I could talk to Pete.

"He's a little busy right now," he said distractedly.

"Maybe you could help me," I asked. "I'm just trying to find out if we're on schedule for taking Sue to NASA."

There was a pause on the other end of the phone. "Sue's not going," he announced. "The F.B.I. is here right now taking her away."

I was getting used to the leisurely pace of photographing wildlife that had been dead for at least 65 million years, but suddenly I was covering fast-breaking news. I was on the next plane to South Dakota. Down at the Black Hills Institute I found not only the F.B.I. but also the National Guard, the Bureau of Indian Affairs, the National Park Service, the Forest Service, the sheriff's officers, local police, and government paleontologists from Dinosaur National Monument and Fossil Butte National Monument, as well as consultants from the South Dakota School of Mines and Technology. They had the Black Hills Institute surrounded as if a live *T. rex* were loose inside the building. Yellow police line tape, the kind they use to seal out spectators from the scene of a violent crime, was strung across the back alley around a storage shed where Sue was kept.

Behind the police tape, a communal rage pulsed through the veins of every Hill City resident as hundreds of schoolchildren and adults screamed protests: "Shame on you—Don't take Sue" and "Don't be cruel—Save Sue." As cars passed the front of the Black

Hills Institute on Main Street, kids held up signs that said HONK FOR SUE. The town appeared to be on the verge of anarchy.

I went across the street to the local drive-in for a Pepsi, and the man behind the counter asked, "F.B.I. or media?"

"Media," I answered. "Why?"

"If you were F.B.I.," he said, "I wouldn't serve ya."

While scores of people protested out front, I went around back and saw the F.B.I. trying to sneak Sue's smaller parts out in boxes through the back door unnoticed. The head of the F.B.I. investigation saw me and said, "Christ, let us do this in peace."

"Maybe you should get another job," I offered, shooting off a few frames of the pudgy chief investigator as he struggled with a particularly heavy box of Sue's bones. I went around the front of the building and whistled for the kids and media to come around the back. As the F.B.I. and National Guard carried off Sue in boxes, kids in the background held placards protesting her removal. The photo opportunity I created made front pages and TV newscasts around the world. So much for unbiased journalism.

Thousands of articles would soon be written about the saga of Sue the tyrannosaur and the private fossil collectors of the Black Hills Institute.

"I'd always been interested in dinosaurs but had never collected one," Pete Larson tells me a few months after the raid. "Back in 1977 we were approached by a museum in Vienna about getting a dinosaur for their new dinosaur hall that they were building. I said, 'Sure we can do that.' It was a lot harder to find a dinosaur than we thought. We started looking in the Lance Creek area, which has been known for years as one of the famous dinosaur localities. We didn't have too much success. Part of it was not really knowing how to look for dinosaurs. We were used to looking for fossils that are smaller, and as they're exposed, the parts have a tendency to remain together. Dinosaurs are a little different. Especially when you look in the Lance Formation, you start looking for dinosaurs and they are weathering out of the hillsides, and as they weather out, they break into half-inch pieces or smaller. If you're expecting to go out looking for dinosaurs and find a whole dinosaur lying out there exposed for you to dig up—forget it, you're not going to find it. We spent two summers looking with little success. We may have actually found something but didn't even know it."

By the spring of 1979 Pete and Neal had put out the word to a lot of their friends that they were looking for a dinosaur.

"A friend of ours was out looking for ammonites," Pete says, "and stopped into a ranch to ask if he could look. The ranch owner was Ruth Mason, who said, 'Well there's no ammonites around here, but there's dinosaurs.' The friend told us about it and we went to the ranch to look around. Dinosaur bones were coming out of the ground all over the place. She had written letters to the School of Mines and to various other schools for years. At that point she was in her eighties, and everybody thought, *Just a crazy lady*, or just plain didn't care—this is a time when dinosaurs were not in vogue."

That part of Mason's ranch is now leased by the Black Hills Institute and is called the Ruth Mason Quarry. They began digging there and found that 99 percent of the fossil material belonged to a well-known dinosaur called *Edmontosaurus annectens*, a hadrosaurian (duckbill) dinosaur that lived at the very end of the Cretaceous, about 65 million years ago. They estimate that there may be as many as twenty thousand individuals in the quarry.

"After thirteen years of digging we still don't know the cause of the herd's death," Pete says. "It was probably a natural disaster which struck this herd. It might have been a flood or a disease, but the catastrophe was localized. Other dinosaurs were still around because some came to feed on the carcasses. We find a few teeth marks on the bones and a lot of shed teeth from these meat-eating dinosaurs like albertosaurs or dromaeosaurs and even *T. rex.*

"The skeletons that we mount are composite skeletons. It takes about five hundred bones to mount a skeleton, and there's no guarantee that even two of the bones in these mounts are from the same individual. One of the complications is that it was a living population—a herd containing individuals which were perhaps forty feet [12 m] long down to individuals which were perhaps fifteen feet [5 m] long. We are currently working on the eighth dinosaur which will be mounted from our quarry. That skeleton will be sold to the University of Tokyo for three hundred fifty thousand dollars. That sounds like a lot of money, but the expense in putting together this elephant-sized animal is sobering.

"Our biggest cost is labor," Pete says. "It takes about fifteen thousand person hours to prepare one skeleton. Now if you calculate that out, that's only twenty-some dollars an hour. We're able to do it at that rate because we get a lot of volunteer help. The digging of the bones itself is not the biggest cost, however. In one quarry season we may get a thousand bones. We have dug over nine thousand bones to date. But because you have a size variation from a twenty-one-inch [53 cm] femur to a forty-eight-inch [122 cm] femur—from an individual fifteen feet [5 m] long to an individual up to forty feet [12 m] long—you've got a real disparity. Not only do you have to find a right and left femur, but they have to be the same size to match with all the other bones. We've now dug ninety-two left femurs. Which means that there's at least ninety-two different individuals that we've dug parts from. It's a real challenge to make sure you put these things together anatomically correctly. You've got to have the right-sized vertebrae, the right-sized lacrimal, the right-sized femur, and the right-sized rib."

Detractors of the commercialization of fossils argue that fossils belong in the country in which they were found. On this argument Larson takes the philosophical high road.

"Fossils are the history of life," Pete explains. "They are the evidence showing the evolution of life on the planet. Fossils in this country don't just belong to people who happen to live in this country now. A child in Japan has just as much right to see a dinosaur fossil as a child in the United States.

"If we've got a quarry that has twenty-some thousand individuals buried, why the hell not spread them all over the world? With twenty thousand dinosaurs in the ground, I don't think it's a real problem to put one in every museum in the world.

"That's the best possible thing we can do. Dinosaurs are a wonderful hook to bring kids into science, and science needs hooks. They start studying dinosaurs and pretty soon they start studying them more in depth—and eventually they branch into physiology or zoology or engineering."

Some paleontologists also argue that private collectors aren't responsible scientists—that in their haste to make a dollar, they'll collect the specimen while neglecting to get other valuable scientific information.

Pete Larson argues, "More than eighty percent of our business is with museums. We *have* to collect proper data or the museums aren't going to want to buy specimens. They are not just looking for display items—they are also looking for items which will enhance and make their collections more important, which make people want to come to see these specimens. So there's a monetary reason for us to collect data. I'm a scientist, too, but just because I'm self-employed rather than working for the government doesn't mean that I'm not going to act responsibly as a scientist."

SUE "THE TYRANT QUEEN"

Indeed, the careful techniques of the Black Hills Institute have paid off and led Pete and his team of scientists to discover more about *T. rex* behavior than has ever been imagined.

"Sue is the finest, largest, the most complete, and the best-preserved *Tyrannosaurus* skeleton that's ever been found," he says. "Her skull is magnificent. Her bones are so well preserved you can see the most minute details of muscle attachment. The preparation techniques which we are using on Sue are far superior to those used on any other *Tyrannosaurus* specimen, and thus we are able to see things that would never have been seen in dinosaur skeletons before.

"We've developed new preparation techniques over the years, and one of those techniques for the final cleaning of the bone is to use fine, dust-sized [50-micron] particles of sodium bicarbonate [baking soda], which are blown through a machine with compressed air at one hundred ninety psi [pounds per square inch]. This doesn't harm the fossil because the particles of sodium bicarbonate are softer than the calcium phosphate which makes up the bone. But they can remove the particulate grains of rock and glue and all the other things that are on the surface of the fossil so that you can get the actual surface of that bone exposed like it's never been exposed before. Terry Wentz, the main preparator of Sue, is a magician when it comes to preparing fossils. His skill is natural, sort of as an artist is born with a skill."

Also contributing to her ideal preservation is that when Sue died, she was buried by sediment so rapidly that there was absolutely no weathering or decomposition of the bones.

The exquisite preservation of this *Tyrannosaurus rex*, which means "tyrant lizard king," led Larson to determine that Sue was probably a "tyrant lizard *queen*."

"It just so happens that Sue probably is a girl," he says. "In predatory birds there are sexual differences in the skeletons. There is a robust and a gracile form. The robust, the larger form, are the females. A researcher quite a few years ago rather chauvinis-

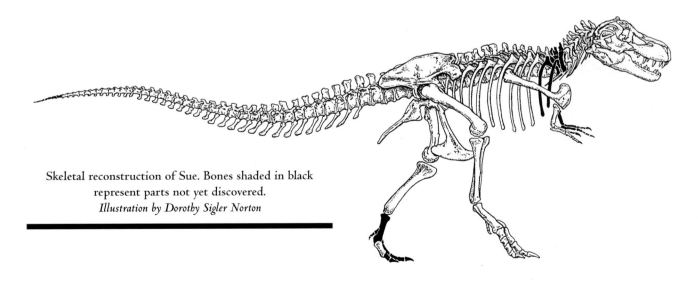

Skeletal reconstruction of Sue. Bones shaded in black represent parts not yet discovered.
Illustration by Dorothy Sigler Norton

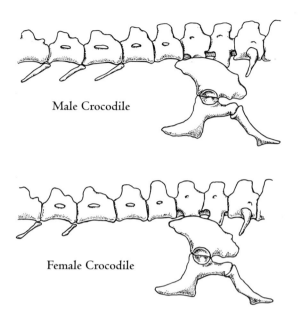

Male Crocodile

Female Crocodile

ABOVE: Sexual differences in caudal vertebrae of crocodiles
Black Hills Institute, after research by Eberhart Freye
BELOW LEFT AND RIGHT: Pete Larson discovered how to tell
the girl and boy *T. rex*es apart. Sexual differences in skeletal
forms of *T. rex* are revealed in pelvis and tail vertebrae;
the gracile form (male) of *T. rex* had an extra larger chevron
at the base of the tail for the penis retractor muscles, the
robust form (female) a smaller one, possibly, Larson thinks,
for easy laying of eggs.
Illustrations by Dorothy Sigler Norton

tically named this 'reverse sexual dimorphism.' We human beings tend to believe that the male should be the larger and stronger of the species, but this is not always true in the animal kingdom. There are many, many different groups of animals in which the female is larger. In invertebrates, almost invariably the female is larger, and also fishes, reptiles, and many birds—but particularly in predatory birds. The most convincing bit of evidence comes from Eberhart Freye in Karlsruhe, Germany, who studies living crocodiles but has a strong interest in dinosaurs. And he's made quite a number of crocodile dissections and noticed that the male crocodiles and female crocodiles differ in the hemal arches at the base of the tail. Hemal arches actually go all the way down the tail, but if you count beginning at the base of the tail, the first and second chevrons in a male crocodile are the same size, and in the female crocodile the first chevron is approximately one-half the size of the second chevron.

"To my knowledge this had not been noticed before in crocodilians. People had never found a way to tell skeletons of crocodiles apart. But this seems to work very well. There's probably two reasons for this difference. One is it would allow the female more space for the eggs to come out past the tail and ischia. And in the male crocodile the penis retractor muscles are actually attached to that first chevron, so it's a good idea to have that first chevron be a little bit more massive to have more room for muscle attachment. In addition we also found that the first chevron

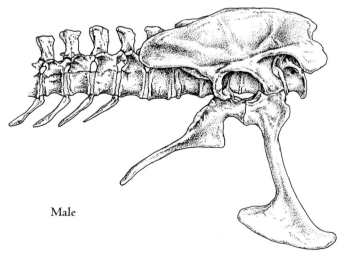

Male

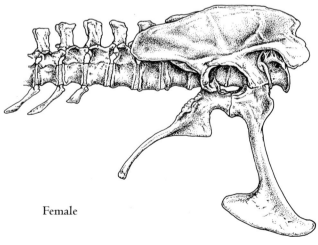

Female

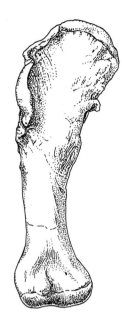

Gracile Morphotype   Robust Morphotype

Differences in male and female _T. rex_ humerus
Illustrations by Dorothy Sigler Norton

is located one vertebra forward on the male! It turns out that this is a very easy way to determine the sex of dinosaurs if that particular portion of the skeleton is preserved. No one has ever really done this sort of thing before—it's exciting."

BAD TO THE BONE

"With Sue there are so many neat things that are preserved," Pete says, "in particular, the evidences of pathology, disease, or injury to Sue's skeleton. Boy, you can take a little instant of time, sixty-five million years ago, and you can see a _snapshot_ of her life.

"For instance, she has a broken fibula. The small bone in the lower leg is smashed. We know that she survived this injury because the bone was healed. When the bone grew back together, it made this huge mass of exostosis, extra bone, that formed over the original shaft of the bone. The nice thing about pathologies, healed injuries, is you can imagine what might have happened to the animal to get them. I believe that this one could have happened in an encounter with an ankylosaur, who in defense smashed Sue in the leg

with his big club tail—the bone break is at the perfect level for an ankylosaur tail to have smashed it or to have been struck by a _Triceratops_."

An injury like that is fatal to almost any animal in the wild—but contemplating Sue's convalescence led Pete Larson to some fascinating conjectures.

"You have to remember, _Tyrannosaurus_ is a bipedal animal," he explains. "It uses two legs to walk on, not four, and here's Sue with a broken leg. How could she feed herself?"

As I ponder the question, Pete fixes his kids some microwave popcorn and continues. "There's a couple things that this brings to mind. Perhaps she eventually won the battle with the ankylosaur and she had food right there with her so she'd have something to sustain her while she went through a healing process. But maybe there's two other possibilities. She was actually able to walk through great pain. Or the other possibility that I like to think about is that maybe she was fed by a mate that brought her food. Here's another aspect of behavior that no one even would have suspected. We're talking about an animal that has remarkable healing properties, which would lend to the belief of warm-bloodedness, too."

Another injury led Pete to establish unequivocally that Sue survived a fight with another _T. rex_.

"We found that one or two ribs were broken and had tremendous massive exostosis—this abnormal bony growth. When we cleaned them, it became obvious that this bone had been infected. And it probably was infected from the time of the injury to the time of Sue's death, because it was never able to heal—it would have been leaking pus. As we cleaned this bone, way in the center we found a piece of a tooth from another _Tyrannosaurus_! Sue was bitten by another _Tyrannosaurus_! Absolutely, positively. And she survived. But she was in pain from that wound for the rest of her life.

"Another interesting injury is her tail, which has forty-six vertebrae. The twenty-fifth and twenty-sixth vertebrae have been smashed. In fact one of them is smashed so badly that part of it is turned sideways and fused together. Sue had her tail stepped on, and I think that happened during mating. Rough sex. Normally when she's fighting, she's so much bigger than other animals and her head would have to be down, so that the tail would be way up in the air where it's not susceptible to injury. The only time it would be susceptible is when she was mounted by a mate. That makes a perfectly good scenario, plus it adds a little bit of sex to the story."

In the past it was thought that the arms of *T. rex* were evolving into uselessness, but another healed injury led Larson to the conclusion that not only were her hands useful, but very powerful.

"There's a muscle attachment on the back of her arm—a tendon on the back of her humerus which was pulled away from the bone, and the bone actually grew around it and reattached that tendon to the bone. This tells us that she used her arms—her arms were not useless—otherwise that muscle would not have been pulled from her arm. Examining her hand, we saw that her two fingers were slightly opposable—she would have been able to use them like ice tongs for grabbing food. I believe that she used her arms to hold onto her prey while she used her mouth to tear out huge pieces of flesh and bone and swallow them—particularly when she made her attack. Maybe when an animal she caught lunged away from her, it pulled this muscle out of her arm."

Pete believes one injury Sue received was her last.

"I believe Sue died as a result of a fight with another *Tyrannosaurus*. We found that the left side of her face had been ripped off. We're not positive about it being another *T. rex*, but the puncture marks across the bone on the side of her face indicate the spacing of another tyrannosaur's teeth. There was nothing else around that could do that unless there's a bigger dinosaur that we haven't found yet, which is another very interesting possibility.

"Sue was partially articulated, although her neck was articulated but broken away from the body—I believe that whoever killed her might have done that. Her head was disarticulated from the neck, although the right side of the jaws was still in place. Either she fell into a streambed after her injuries, or just shortly after she died a flood came along and buried her, which means the aerobic bacteria were not able to act upon her skeleton as the flesh decomposed.

"We found food still within her digestive tract. And very nearby we found acid-etched bones that I believe are the remains of her last meal—an *Edmontosaurus annectens.*"

ROYAL FLUSH: THE TYRANT QUEEN'S FAMILY?

But probably most surprising is that with Sue they found parts of three other tyrannosaurs.

"Next to Sue we found the end of a tibia and fibula from a smaller male tyrannosaur, probably half the mass of Sue," Pete says. "Sue stood thirteen feet [4 m] at the hips and was approximately forty-one feet [12.5 m] in length. Her head was five feet [1.5 m] long; when she stood up, her head could have easily been twenty feet [6.1 m] above the ground. She could have weighed as much as eight tons. The smaller *Tyrannosaurus* probably would have been only ten feet [3 m] tall at the hips, maybe even a little bit shorter. Its tibia and fibula were bitten in half. We know they were bitten in half because of the way the tibia is broken and the marks which were found on the tibia. And on the fibula itself we find marks which show tooth serrations from another *Tyrannosaurus's* biting it. But in addition to that specimen we found the lacrimal, which is a bone in front of the eye, from another, smaller *Tyrannosaurus*. Its skull would only have been about eighteen inches [46 cm] long. But, even more surprising than that, we found the frontal, a skull bone which covers the roof of the brain, from a fourth *Tyrannosaurus*. This *Tyrannosaurus* would have had a skull about ten inches [25 cm] long—a baby *Tyrannosaurus*. The baby would maybe only have weighed a couple hundred pounds [91 kg]—Phil Currie calculated that this baby *T. rex* was six feet, eight inches [2 m] in length and about thirty inches [76 cm] in height.

"Perhaps it was Mom, Dad, Junior, and Baby," Larson says. "Very, *very* strange. No one would have ever suspected that they potentially moved in family groups. They've always been thought of as being lone travelers."

Pete believes that Sue was just beginning to tell her 65-million-year-old secrets when the tyrannosaur's life got complicated by a modern nightmare—litigation.

"We excavated Sue by hand in seventeen days," he says. It was an ordeal that landed him on his back for days and rendered his hands without feeling for weeks. "When Maurice Williams, the landowner, first invited us

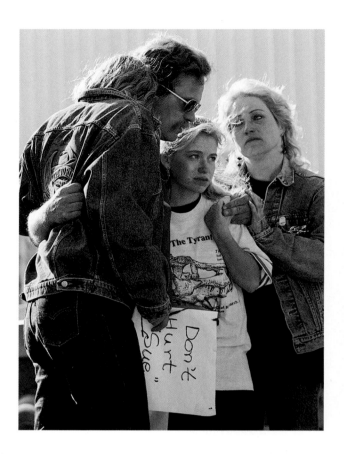

to look for dinosaurs and collect stuff on his property, I said, 'You know if we find something good, we'd be happy to pay you a little bit of money. We don't pay a lot, but my dad was a rancher and I know how much work it is, and how little income one gets out of his ranch.' He said, 'Oh no, that won't be necessary. The only thing I ask is that you do not drive on the property. You just walk and carry things off, don't drive on and pick them up.' So when we found Sue, I called Maurice up and told him that it's something big and it looks like it's going to be something good. I invited him over to show him where it was in the cliff face.

"After we called him, we then called up the courthouse and found out that, yes, indeed, he did own the property and that he even had mineral rights, because Amoco Oil had an oil lease on there for something like twelve years. He leased out the grazing rights, too. He told us he bought the land in the 1960s. About half the land on the reservation is privately owned. I paid him five thousand dollars to collect Sue. It's still the most that has ever been paid for a fossil still in the ground. That five thousand dollars is almost as much as that quarter section of property is worth. It's one hundred sixty acres [64.8 ha], a half mile [0.8 k] by a half mile [0.8 k], and up there land goes for about fifty, maybe sixty, dollars an acre.

"We found out later that the land title was in trust—but not from Maurice. We got this weird phone call from the F.B.I. wanting to come and see the dinosaur and I said, 'Sure, come on up and talk to us, no problem.' But then my lawyer called them up and found out we were the target of an investigation. When that happened, we did a real close check and got a copy of the trust document. It said, 'The United States of America shall hold this land in trust for the sole use and benefit of Maurice Williams and his heirs for twenty-five years.' That doesn't mean it belongs to the United States government. The only reason to put land in trust is to avoid paying taxes. If you put money in the bank for a trust for your children, the banker or your brother or your wife might be the trustee, but the trustee would not own that money! They are only looking after it for that person. The term 'trust' doesn't mean they have ownership."

To put salt on an old wound, advising the F.B.I. at the raid was Pete's old master's adviser. The thirty F.B.I. agents who descended on the Black Hills Institute during an early morning raid on May 14, 1992, thought they would cart Sue off in the trunks of their cars. They had come prepared to take her with a few cardboard boxes, a couple of rolls of duct tape, and a roll of bubble wrap. Even with the help of a Black Hills Institute crew, it took the F.B.I. three days to pack up Sue. The T. rex's crates would eventually fill two forty-foot-long [12 m] semitrailers.

When the National Guard, dressed in battle fatigues, carted Sue away, the townspeople were locked together, hand in hand, forming a human chain lining either side of the road in back of the Black Hills Institute. Tears were falling from old men comforting their grandsons; wives were comforting their husbands who were comforting their children. Sue the tyrannosaur had died 65 million years ago, but she had awakened the hopes and dreams of the community, which guarded its effort to start a museum that could revitalize the town.

"The dream of the museum is not just our dream," Pete tells me later. "It's the dream of the people in Hill City, too. In fact, the kids were raising money before they seized Sue. A week before the raid, the fourth-grade class gave a check for the sixty-one dollars and twenty-eight cents that they had raised in a bake sale to the museum. When the F.B.I. came in and seized Sue, it was not just us who felt raped; the entire city and the entire area felt raped.

"When Sue was being prepared, she had more than two thousand visitors. Most of those were kids coming down to watch Sue being prepared. Everybody knew about the museum that was being built. Everybody became personally attached.

"Those of us who grew to know Sue also grew to love her. She was like a person to us. She's an entity. She's not just a bag of bones. She's not just a pile of rocks. She was a living, breathing thing that was coming to life for us.

"As we were uncovering the mysteries, you could see little glimpses of her life. Sue has personality. Whenever I felt bad, I'd go out and look at Sue's skull, and it would make me feel so good to see that wonderful, wonderful fossil. And I think that just about the entire town felt the same way. This wasn't just our museum. This is the town's museum. This is a community's museum. It's something that got everyone involved.

"And this is what's so asinine about this whole situation. Kevin Schieffer, the acting U.S. attorney, says, 'It's property of the United States government. Period.' You know he's acting U.S. attorney because he wears makeup to raids. He called the television station up, put makeup on, and seized Sue and all our records. And then they subpoenaed us again and took fifty thousand more documents. They went through our whole place. They could have taken whatever they wanted and they did. They even took things like our lawyer's file and cleared off our order board with all the orders we had. So we don't even know what orders we had there. Just all kinds of things to disrupt our business—harassing all our friends. Trying to get them to tell them something that we have done wrong. They even interviewed my high school teacher, whom I haven't seen in twenty years. They've convened two grand juries and they haven't brought an indictment against us.

"The entire science of paleontology receives about a million dollars' worth of grants from the National Science Foundation a year. That's about all the money that is allocated to paleontology. That's the entire field of paleontology. That's invertebrate paleontology, too. The people who study clams, the people who study foraminifera, people who study trilobites, people who study mammal ears, people who study rodent teeth, people who study dinosaurs. All that money is split up among all these paleontologists. About one million dollars a year. This investigation has already cost the federal government more than a million dollars. We've spent more than seventy-five thousand dollars to date on legal fees, and I don't know how much it's going to end up being by the time this is all over with. [It stood at $250,000 one year later, in October 1993.] If you used that money for paleontology, for science, just think of the wonderful things you could do. You double the amount available to paleontologists every year."

"Amateur collectors make the big discoveries," Pete tells me. Susan Hendrickson, who found Sue, was an amateur. Stan Sacrison found another of our tyrannosaurs—he's an amateur, too."

The tyrannosaur called "Stan," again named after its discoverer, was found shortly after Sue. Stan was about 60 percent complete.

"We just got Stan the *T. rex* in 1992," Pete says, "but Stan actually found him five years before. He found an articulated vertebrae of a dinosaur and a pelvis sticking out of the ground, made an excavation of part of it, and took my old master's adviser from the School of Mines and Technology to see it. He identified it as just another *Triceratops*! Now there are only three mounted *Triceratops* skeletons in the world, but this sort of shows the way some people think. It looked like a lot of work. You gotta remember, digging dinosaurs is a lot of work. The dinosaur just doesn't crawl out of the ground and into the back of your truck. So he just sort of left it alone and didn't think about it for five years—and discouraged Stan from collecting it. Well, if you have just three articulated *Triceratops* in the world, holy shit! Go out and collect that sucker. Don't leave it to lay in the ground and rot. If you've got an articulated skeleton, you have a moral obligation as a paleontologist to go and collect it.

"A mutual friend said he was up hunting with Stan and he showed him a skeleton that he had found before, and there was an eight-foot [2.4 m] section of vertebrae exposed. And I said, 'Really!' This was in January. And I said, 'When can we go? Can we go tomorrow?' And he said, 'Well, I'll give him a call.' So we went up to look at it. I was ten feet [3 m] away, and I sat down and said, 'Holy shit! Another *Tyrannosaurus rex!*'"

Coincidentally, it turned out that Stan was probably a male *T. rex*.

Just after the raid, I was down in the offices of the Black Hills Institute with Pete and Neal Larson. We all had a hollow feeling inside, as if someone close to us had died, but Pete Larson, never one to stay depressed very long, began to cheer everyone up.

"We feel pretty bad about losing Sue," he said, "but what about Stan? After sixty-five million years he found the perfect woman and now she's gone."

His brother Neal, eyes swollen from tears, began to smile. "Stan was too young," he replied, "a thousand generations too young."

"Sue liked younger men," said Terry, the preparator who spent over 2,300 hours preparing the block of stone from which her skull emerged.

A journalist came by and said, "Got anything quotable for me?"

"Yeah," said Neal. "You can say that the F.B.I.'s tyrannosaur adviser up there wouldn't know a *T. rex* from a *Triceratops* if he saw one lying in front of him."

SIOUX SUE FOR SUE

It has been two years now since the seizure of Sue, and the most recent court—not being able to cite a legal precedent for fossils—declared Sue to be real estate. And since reservation land needs government approval before it is sold, the court deemed that the transaction to the Black Hills Institute was null and void—and awarded the 65-million-year-old parcel of land called Sue to Maurice Williams, the landowner. But now the Sioux Nation is suing to get Sue. They reason that since Maurice Williams didn't apply for a hundred-dollar peddler's permit to sell Sue, she now belongs to the Sioux.

But the court hasn't heard the last from the Black Hills Institute.

"I'm not giving up if I live to be eighty," Pete, who is forty-two, tells me. "It's simply not right. We'll take this all the way to the Supreme Court if we have to."

In the summer of 1993 Stan Sacrison discovered his second *T. rex*, the third one to be collected by the Black Hills Institute. The crew named that tyrannosaur "Duffy," after their tenacious lawyer Patrick Duffy. The Sue custody case is still on appeal, and for now she remains under the protection of the court—in a shipping container at the School of Mines in Rapid City, South Dakota.

EXTINCTION: THE FINAL CHAPTER

Our obsession with dinosaurs is closely linked with why they disappeared. Dinosaurs were the biggest, some of the most diverse, and the longest-living land animals, and suddenly, remarkably, they were gone. Humans are much like dinosaurs. Even though we lack the diversity and longevity of the former tenants of the earth, we are likewise at the top of the food chain, we cover the entire globe, and we are, right now anyway, seemingly invincible.

The great Harvard naturalist E. O. Wilson, in his book *Diversity of Life*, tells of the five great extinctions of the last half billion years and delivers a sobering chronicle of the cataclysmic extinction we are in the middle of right now. But this time there isn't a doomsday comet to blame it on, or the deadly irradiating rays of a death star, or a rogue cloud of asteroids from the great void to lay waste our fragile blue oasis. This widespread doom is being inflicted by a single agent, our own species, *Homo sapiens*, ironically the one animal that uses its brain size to distinguish itself from the other animals.

Without a well-developed, compassionate heart to guide it, the brain is simply a container for a lot of information adrift in a tangle of cells.

During the human reign, species are being lost faster than we can document them. As we encroach on every available niche, recklessly polluting the atmosphere, the ocean, and the land, we are doing what no animal in the wild would do—foul its own nest.

Evolution of the heart will be our salvation, and it starts with knowing from where we have come. Perhaps that is the biggest value of dinosaurs. Among a multitude of lessons they can teach are that brain size isn't all that important, big isn't always better, change is good, and extinction is forever.

For all the beauty of the earth and its remarkable ability to sustain life despite adversity, there is no colder heart than Nature when the guests have overstayed their welcome.

The last of the great pterosaurs, *Quetzalcoatlus*, is silhouetted against the flash of a meteor impact along the Yucatán coast some 65 million years ago.
Painting by Douglas Henderson, from Dinosaurs: A Global View

PICKING UP
THE PIECES

THE DINOSAUROID

Dale Russell, curator of fossil vertebrates at the Canadian Museum of Nature in Ottawa, thinks that toward the end of their demise, dinosaurs were showing a predisposition for humanlike features. Using extrapolated data, he created a creature called the Dinosauroid, which is his best guess as to what one of the smarter dinosaurs would look like if it lived today.

"What happened originally was in the American Museum of Natural History I saw a skullcap from the scraps of a dinosaur and I thought, *Wow, here's an animal that had a really large brain for a dinosaur.* Nobody had picked up on that before. So I went back to Dinosaur Park and I spent a month, six weeks, just beating myself to a pulp on the outcrop trying to find some remains of that dinosaur. I didn't find any. And then an Alberta rancher's wife took me to a place where she saw some remains that I recognized as this beast, and I got another skullcap and a fragmentary skeleton. It's still the most complete skeleton of a *Troodon* in North America. So we collected it, and when I was working on it and putting it together, I realized, *Well, isn't this interesting.* It had big eyes. They were somewhat focused ahead. I had its wrist, and I knew that it had a tridactyl hand. Obviously it had a large brain, and it's bipedal, and what does this remind you of?

"If you take the ancestors of man, the little quadrupedal, naked, tailed, hairy beasts around at the Late Cretaceous that we are descended from, they

ABOVE: Ottawa, Ontario
OPPOSITE: Dale Russell thinks that toward the end of their demise, dinosaurs were showing a predisposition for humanlike features, bigger brains, and more grasping hands. This led him to create the Dinosauroid, a model of what he thinks one of the smarter dinosaurs, like *Troodon*, would look like if it survived until today.

look less like us than some of the small theropod dinosaurs. Anyway, Richard Leakey came to town and gave a talk, so I took those bones and I said, 'Gee, you know, Dr. Leakey, sir, you should have a look at these bones.' He looked at them and they amused him. He didn't say anything and he was smiling and his eyes were smiling, too. Because the completeness of that specimen was similar to the completeness of hominid specimens in East Africa. I think he was having a sweet and sour experience looking because it was so different and so similar at the same time.

"We made a model of the *Troodon* skeleton, and the model maker, Ron Seguin, had a little time left over, so I thought, just as a gas, why don't we try to extrapolate the encephalization curve that I had already worked on with correspondents with the Search for Extraterrestrial Intelligence people at NASA. The curve is so interesting because it says that, through time, the probability of human brain-body proportions are rising. . . . They are measurable and quantifiable. And Ron thought, *Well, let's fit this dinosaur into the curve and try to think in terms of functional anatomy and how to support it with the curve.* So we started with the head, and then worked down from that."

Russell's provocative sculpture has been the subject of much controversy because most scientists think of evolution as random, not progressing, as Russell thinks, by quantifiable improvements.

"If you look at mammal skeletons, they are beautiful things," he says. "They have so much complexity compared to dinosaurs. Dinosaur skeletons are so crude, by comparison—they're wonderful in the proper perspective. But my acquaintance with the dinosaurian world just makes me marvel at the modern world. Having studied dinosaurs for thirty years, more or less, I just look at the modern world like a miracle— it is so interesting and so rich, and man is the same way, because . . . How can I say this? Man is extraordinarily successful."

DINOSAURS IN WARTIME: THE SECOND EXTINCTION

World Wars I and II were disastrous for many dinosaur families, many of whom suffered a second extinction as innocent bystanders to human conflicts.

The Sternbergs were a family of dinosaur collectors from the United States and Canada who sold specimens to great museums around the world. They had collected two duckbill dinosaurs called corythosaurs from the Dinosaur Park Formation in Alberta,

BELOW: *Spinosaurus,* a strange sail-backed dinosaur, was among many unique dinosaurs that faced a second extinction during the World Wars.
Illustration by Michael Trcic
OPPOSITE: Dinosaurs lost in wartime
Illustration by Pat Redman

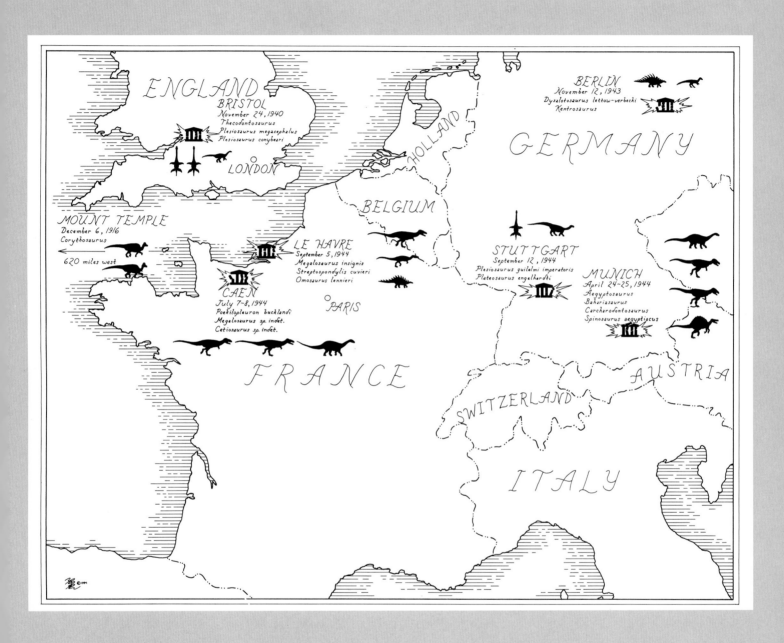

ENGLAND

BRISTOL
November 24, 1940
Thecodontosaurus
Plesiosaurus megacephalus
Plesiosaurus conybeari

LONDON

MOUNT TEMPLE
December 6, 1916
Corythosaurus

620 miles west

LE HAVRE
September 5, 1944
Megalosaurus insignis
Streptospondylis cuvieri
Omosaurus lennieri

CAEN
July 7-8, 1944
Poekilopleuron bucklandi
Megalosaurus sp. indet.
Cetiosaurus sp. indet.

PARIS

FRANCE

HOLLAND

BELGIUM

BERLIN
November 12, 1943
Dysalotosaurus lettow-verbecki
Kentrosaurus

GERMANY

STUTTGART
September 12, 1944
Plesiosaurus guilelmi imperatoris
Plateosaurus engelhardti

MUNICH
April 24-25, 1944
Aegyptosaurus
Bahariasaurus
Carcharodontosaurus
Spinosaurus aegyptiacus

SWITZERLAND

AUSTRIA

ITALY

Canada, one with rare skin impressions that was said by Charles H. Sternberg to be one of the finest they ever collected. The pair of duckbills were loaded aboard the English steamer *Mount Temple* for their journey to the Natural History Museum in London via Brest, France. But the specimens never arrived. On December 6, 1916, the boat was blown up by the German raider *Moewe*, a warship disguised as a common cargo vessel, and the specimens are still sitting at the bottom of the Atlantic 620 miles (998 k) west of England.

It was World War II, however, that brought the most destructive force upon the Dinosauria since the Cretaceous extinctions.

At the outset of trouble, the British Museum of Natural History in London sent its dinosaurs to wait out the war in the chalk caves of Kent. Those dinosaurs remained safe, but on November 24, 1940, a German air blitz dealt a blow to Britain's aerospace industry in Bristol, and in the process wiped out the Bristol City Museum, decimating an extraordinary collection of plesiosaurs as well as some material from one of the first dinosaurs ever discovered, *Thecodontosaurus*, a prosauropod described by the curator Samuel Stutchbury in 1836. The plesiosaurs were both type specimens: *Plesiosaurus megacephalus* (Stutchbury 1846) and *Plesiosaurus conybeari* (Sollas 1881).

Allied bombing raids on September 3 and 4, 1943, shattered the great skylights above the *Brachiosaurus* at the Natural History Museum at Humboldt University in Berlin and convinced its curators to dismantle their famous specimen, the largest mounted dinosaur in the world, and retire it to the safety of the basement. It was a wise move. A little over two months later, on November 12, 1943, fires caused by Allied bombing raids completely destroyed the museum's Reptile Hall, cremating the type specimen of *Dysalotosaurus lettow-verbecki* (Virchow 1919), the "uncatchable lizard," along with examples of its Late Jurassic neighbor *Kentrosaurus*, a well-known stegosaur from Tendaguru, Tanzania.

On the night of April 24, 1944, several unique dinosaurs were bombed back to oblivion by a Royal Air Force bombardment of the Bavarian State Collections for Paleontology and Historical Geology in Munich. These dinosaurs, all type specimens, were collected by Ernst Stromer von Reichenbach from the Cretaceous beds of the Baharija Oasis in Central Egypt. They included a sauropod called *Aegyptosaurus* (Stromer 1932) and three large, almost *T. rex*–sized carnivores called *Carcharodontosaurus* (Stromer 1931), *Bahariasaurus* (Stromer 1934), and *Spinosaurus* (Stromer 1915). *Carcharodontosaurus* was named after the great white shark *Carcharodon*, whose teeth it resembled, but the most serious loss was perhaps Stromer's *Spinosaurus*, the only known specimen of a strange sail-backed carnivorous dinosaur.

During the invasion of Normandy, continuous Royal Air Force firebombings on the night of July 7, 1944, incinerated the University Museum in Caen, France. Two megalosaurs perished. One of them was unidentified, but the other, *Poekilopleuron bucklandii* (Eudes-Deslongchamps 1838), was one of the best Middle Jurassic meat-eaters ever found. Also lost in the raid was an unidentified dinosaur thought to be a sauropod called *Cetiosaurus*.

On September 5, 1944, during the liberation of Le Havre, a harbor town at the mouth of the Seine, a fine collection of Jurassic dinosaurs was firebombed during another Royal Air Force night raid. While trying to destroy German bunkers along the coast, bombs rained down on the town's Museum of Natural History, blasting two theropods, *Megalosaurus insignis* (Eudes-Deslongchamps & Lennier 1867) and *Streptospondylus cuvieri* (Von Meyer 1830), as well as the fine stegosaur *Dacentrurus*, making World War II particularly devastating to the stegosaur family.

On the night of September 12, 1944, two hundred Royal Air Force bombers dropped their hellish payload on Stuttgart, destroying much of the State Museum for Natural History and annihilating one of the only complete specimens of *Plateosaurus engelhardti* from the tip of its nose to the tip of its tail. Also destroyed was the museum's crown jewel, the type specimen of *Plesiosaurus guilelmi imperatoris*.

The devastation the World Wars have wrought on our natural history heritage has encouraged many museums to cast and disperse multiple examples of precious specimens to other repositories in case tragedy should ever strike again.

Extinction: What Killed the Dinosaurs?

OR,
The Worst Weekend in the History of the World

Paleontologists have been pondering the demise of the dinosaurs almost since they were discovered. There have been well over a hundred different theories advanced to explain this extinction at the end of the Cretaceous, 65 million years ago, so to bring resolution to this age-old question, we decided to conduct our own survey of the paleontologists we met and asked them, "What killed the dinosaurs?" They seemed to shake out into two basic camps: the catastrophists and the gradualists.

The catastrophists' usual suspect today is a meteor called Chicxulub that created a 150-mile-wide (241 k) crater near the Yucatán Peninsula in Mexico. Recent dating put the impact of this meteor at 65.7 million years ago, too close to the scene of the crime for some to call it a coincidence. But the "impact theory," as it is called, might have had an accomplice in the Deccan Traps—a colossal volcanic eruption that occurred at more or less the same time. This massive volcano vomited so much basalt lava that it formed the highlands of India, and it might have issued enough ash to choke the atmosphere and block out the sun, wreaking havoc on the plants and large animals of the time.

The gradualists, for their part, favor a slow death for the dinosaurs and usually cite the change of climate brought on by continental drift. Volcanic eruptions, rampant spreading of deserts, encroachment of polar ice caps, rerouted ocean currents, the draining of the inland seas, were all byproducts of continental drift, and as "Dinosaur Jim" Jensen told me, "Continental drift can be used to explain everything—from lousy weather to Republicans."

There are a fair number of experts who subscribe to both the gradualist and catastrophist theories and call the culmination of these unfortunate events at the end of the Cretaceous "the worst weekend in the history of the world." This geological equivalent of a really bad day is what I believe did the big guys in.

Here, in alphabetical order, are the results of our survey of what the experts think.

Bob Bakker "We'll never understand the terminal Cretaceous extinctions until we look at all extinctions. It's not one crime of passion in the ecosystem that killed off the dinosaurs. It was this serial killer who operated in the exact same way again and again and again and again."

Rinchen Barsbold "I think the conclusion depends on the temperament of the scientist. I belong to the dinosaurologists who have a more or less quiet temperament and consider the extinction was gradual."

Michael Brett-Surman "Well, humans will eventually go extinct, but the key question is to leave a descendant. I mean, if you evolve into something else, technically you're extinct, but you're still there at the same time. It's like a lawyer's loophole. So now, dinosaurs left behind birds. So meat-eating dinosaurs are still here."

Sankar Chatterjee "Some catastrophists think the meteor impact caused the extinction—the other group believes it is the Deccan Traps that caused it. Both groups are right. There are two monsterites that struck the planet about the same time. One is, of course, off the coast of the Yucatán Peninsula in Mexico. The other is our crater, which we named Shiva, the god of destruction, which triggered the lava flows of the Deccan Traps. Like the moon, the crater is filled with lava."

Karen Chin "I think it was a combination of factors. There probably was an asteroid impact sixty-five million years ago. But I'm of the it's-probably-more-complex-than-we-want-it-to-be school of thought. Dinosaurs were declining in diversity and getting more specialized—the more specialized you are, the more susceptible you are to extinction."

Ned Colbert "It's very hard to say. If it was a meteor, why didn't it kill turtles, crocodiles, birds, and mammals? The meteor would have to have been a smart bomb."

PHIL CURRIE "I lean a little more toward the gradualistic end. Reason for that is here in Alberta, as you go through the last ten million years along the Red Deer River, from seventy-five million years through the end of the Cretaceous, you see a gradual reduction in numbers of species of dinosaurs, although evidence for something major happening at the end of the Cretaceous is also pretty robust."

STEPHEN CZERKAS "Bad press. Undeniably, it had to be environmental changes. A comet might have slammed into the globe approximately the same time as the end of the dinosaurs, but that doesn't mean that it was the bullet that actually killed them."

DONG ZHIMING "At the end they were at their peak of diversity and the mass extinction was a sudden event. Perhaps a coming together of all events may have directly or indirectly played a part. Unfortunately there is no way to know."

JIM FARLOW "I don't know. Although I have to admit that I am increasingly coming around to the idea that maybe an asteroid had a lot to do with it. The dates they're coming up with for an impact are getting pretty darn close for coincidence. The dinosaurs may have been declining before that, but I wouldn't be surprised if this thing took the rest of them out. So I guess I am becoming a catastrophist in my old age. It appeals more to my sense of the dramatic, too.

"If you want to see how mass extinction works, stay tuned. We are probably going to have one over the next generation with destroying the tropical rain forests and our impact on tropical reefs. We are certainly affecting the environment in a rather more significant way, I suspect, than even an asteroid could."

PETER GALTON "I guess I would make it a compromise. I think the dinosaurs were on the way out anyway, and if there was any extraterrestrial event, then that was just the last straw."

JACQUES GAUTHIER "The dinosaurs weren't killed off. They became birds. And if Caesar didn't cross the Mississippi, you don't have to think up a reason for

why he did. Some dinosaurs became extinct at the end of the Cretaceous, just the way that some mammals did, some lizards did, some turtles did, but we don't say that any of those groups are extinct. We only say the dinosaurs are extinct. I put birds with dinosaurs.

"What I worry about is that we are in the process of a megafaunal extinction right now. You go down to the L.A. La Brea Tar Pits and there's elephants and lions, tigers, bears, all kinds of stuff, giraffes, hyenas, cheetahs—all the stuff you see in Africa today was living in L.A. ten thousand years ago. How come it's completely gone out of the entire Northern Hemisphere? The only megafauna left is down in Africa now, and we're doing our best to wipe that out. We haven't really found an altogether satisfactory explanation for that, and that happened right in front of us, so to speak. So now we're supposed to go back sixty-five million years and figure out why the big animals back then bought the ranch? I think that's a really tough one."

DAVE GILLETTE "I don't have a notion. I don't fully believe the single-cause theories. I really haven't done any work at the Cretaceous/Tertiary boundary myself, so all I can do is read what others have done. I'm an empiricist. I don't develop an opinion unless I've got direct experience myself."

KARL HIRSCH "A meteor impact or volcanic eruption could change the whole climate. In breeding, temperature change is critical. If the weather changed for a hundred years, maybe their eggs couldn't hatch. A half-degree change in temperature can kill an embryo."

JACK HORNER "A comet killed the dinosaurs—all two of them. If not, it was old age. When the last dinosaur died of old age, that was the extinction of dinosaurs. There wasn't anyone else there to produce with."

JIM JENSEN "Why they died, I don't know. They were very successful—they filled every environmental niche. I have no theory. I'll leave it to the armchair specialists because it's being dealt with continually by them."

ROLF JOHNSON "We did a big research project here at the museum, trying to look at whether or not the dinosaurs in the Hell Creek were slowly petering out. It looks, from a preliminary reading of the data, as though dinosaur abundance and diversity was reasonably stable there. So I'm a real big fan of an asteroid impact at the end of the Cretaceous. I don't think the data is there to actually pin this down as a cause and effect. Undoubtedly, other things were going on globally at the end of the Cretaceous which might have factored into the demise of the dinosaurs."

JIM KIRKLAND "There's no question the asteroid hit. I don't care what anybody says, and I believe they have the smoking gun in the Yucatán, although I personally believe there's multiple impacts. I was working on the Brazos River when the first tidal wave impact from the meteor was recognized. We have records of a tidal wave over the entire Gulf-Atlantic Seaboard. The wave must have been two to three miles [3.2–4.8 k] high before it hit the continental shelf, laying bare the Gulf at that time. It dates real well, too, and it ties to the marine extinction, where we lose ninety percent of all the calcareous phytoplankton of the world's oceans."

SERGEI KURZANOV "Some events, like an impact and the dramatic change of climate which followed, could have speeded up the senescence of dinosaurs, but in no case such events could be considered as a reason alone for dinosaurs' death. Dinosaurs had a high potential mechanism of survival. They could stand temperatures as low as minus four or five degrees Centigrade as minimum, and some dinosaurs could live in the conditions of polar nights."

NEAL LARSON "They got to be too social. Continents collided together, and these big animals could go from one continent to another and they spread dinosaur diseases like AIDS."

PETE LARSON "The only two theories which are based on geologic evidence are the volcanic eruption and the impact theory where there's one or more potential craters. Perhaps a swarm of comets or asteroids struck the earth. It's a very likely thing to have happened— just look at the moon."

GIUSEPPE LEONARDI "I believe it was gradual. Dinosaurs were becoming more specialized, like for example the koala today who can eat just one species of eucalyptus or the panda who can eat just bamboo shoots. And then we must say also that not all dinosaurs became extinct, because birds are dinosaurs—glorified dinosaurs."

GIANCARLO LIGABUE "All people ask why they disappeared. My question is different. Why did they live so long?"

MARTIN LOCKLEY "It's looking like an impact more and more. In the last six months or so [October 1993] we found dinosaur tracks in at least four levels within two meters [2.2 y] below the iridium layer. Complete skeletons are almost nonexistent in the last three meters [3.3 y] before the iridium layer, but we don't seem to have a shortage of tracks. One layer is thirty-seven centimeters [14.6 in] below the iridium layer. So this shows that in at least the area around southern Colorado around Trinidad, where we've been working, dinosaurs like hadrosaurs and large ceratopsians were living up until the last minute of the Cretaceous."

JACK MCINTOSH "Here is one place I refuse to be quoted. However, I'll tell you what happened in Drumheller when the Tyrrell Museum first opened and Phil Currie got together fifty-two so-called dinosaur experts. They decided to have a secret poll to see what they thought of extinction, and of thirty-six people willing to express an opinion, thirty-two voted in favor of the gradual extinction and four in favor of the catastrophic extinction. Okay?"

ANGELA MILNER "I used to be a gradualist, but I'm more agreeable toward a catastrophist because I think there is a lot of overwhelming evidence that some catastrophic event happened at the end of the Cretaceous. It's something that affected land animals and affected the marine chain as well. I'm more of a volcanist than a meteorologist, as far as this is concerned."

MARK NORELL "First of all, dinosaurs didn't go extinct, because birds are still with us. Birds are more closely related to *Tyrannosaurus rex* than *Tyrannosaurus rex* is to

Triceratops . . . only some dinosaurs became extinct.

"The evidence for extraterrestrial impact is pretty equivocal, and as far as sudden or gradual depends on how you measure time. What do we consider sudden? It's before lunch or before dinner, but the best resolution we have to measure geologic time is ten thousand years, and beds where dinosaurs are found are more like one hundred thousand years. In the last one hundred thousand years we've had several major glaciations that almost covered the entire planet, we've had the emergence of human culture, we've had the extinction of great faunas, and we look at the processes as being pretty gradual. I'm not willing to go with that being sudden.

"If you're going to extrapolate from this huge meteorite hitting at the end of the Cretaceous to being the final and ultimate cause for dinosaur extinction, it can't be supported by the evidence. Empirically we can't really test these things and it seems kind of a waste of time to sit and write scientific papers about this."

JOHN OSTROM "You may have an impact date at almost exactly the right time. It's close enough, so that's probably the site that produced the iridium spike, and that may be a worldwide event, and it almost certainly had a contribution to the change in the fauna, but I still worry about all those things that seem not to have been upset."

KEVIN PADIAN "I think that it's environmental. It wouldn't surprise me if we find much stronger evidence that the climate was deteriorating for them. Dinosaurs have always had a very high turnover—it was a way of life for these animals; you almost get a hundred percent turnover for every formation. And toward the end of the Cretaceous they weren't speciating as rapidly. I think we can stop looking for dinosaurs doing business as usual and suddenly the asteroid hits. Nonsense. Absolute nonsense."

ARMAND DE RICQLÈS "There are several possibilities, extraterrestrial things, this is one. Volcanism, which is another. Change of climates. The collapse of the various ecosystems due to the marine regression at the end of the Cretaceous. All explanations may hold a part. When somebody says this is the cause and this the effect—the dinosaurs died—I think this is a naïve approach. Because in all historical sciences we see that the causation is a complex one and it's circumstantial causation, which probably comes from the fortuitous linkage of several factors which incidentally happen to occur more or less at the same time. You cannot say that the Second World war was caused by precisely one thing."

DALE RUSSELL "You cannot study why the dinosaurs died by studying dinosaurs. It's just crazy. It's insane. There are three hundred eleven skeletal fragments of dinosaurs, worldwide, for the last nine million years of Cretaceous time, and you have billions of atomic nuclei which you can examine, like pollen grain, soot, and shells. Why would you overlook a body of information that gets you resolution like that? I have zero patience for guys who are going to prove that dinosaurs are dying out slowly by counting dinosaur skeletons."

PAUL SERENO "For the dinosaurs, I think climatic factors played a bigger role than the gradual ones. The seas dropped quite rapidly at the end of the Cretaceous and caused a whole series of climate changes. In less than a million years a seaway that stretched from basically Pennsylvania to Colorado dried up and drastically changed the climate that the animals strolling along the beach had to encounter. I do believe however that there was an asteroid impact that may have caused the final crisis, but I think climate had more to do with it."

PHILIPPE TAQUET "I don't believe in catastrophic events. Not at all. I am a gradualist. I think, like Sullivan, who wrote in '87 that there were several extinctions of different lines of reptiles and dinosaurs beginning at the end of the Campanian, eight million years before the Cretaceous/Tertiary limit—not one at the end of the Cretaceous."

DAVID WEISHAMPEL "There is a decline in hadrosaur diversity toward the end of the Cretaceous, so that by the time you are at the very, very end, there were probably two, maybe three kinds of hadrosaurs that were looking up into the sky watching an asteroid plunging into the atmosphere."

SAM WELLES "I don't know enough about the dinosaurs at the end to answer you. My feeling, which is not based on accurate work, is that they were thinning out toward the end. There seems to be pretty good evidence that there was an impact that might have been fatal to large animals. The only objection that I have is that a lot of things came through. Mammals were contemporary with the dinosaurs and expanded rapidly. But why weren't the crocs killed off? Why weren't turtles killed off? Why weren't birds killed off? There are unanswered questions, so I think it's still open for more evidence."

EPILOGUE: THE DINOSAUR ROAD WARRIOR

The thirty-three-foot-long (10 m) fiberglass *Allosaurus* was roped to the back of Albuquerque sculptor Dave Thomas's pickup truck, and as he drove down the road, the dinosaur was shaking, looking more like it was being restrained than supported. Dave was taking the beast to California, whence it was going to be shipped to a museum in Japan. We decided to tag along in another vehicle and photograph this strange migration.

Dave had stopped to refuel at a gas station near the Zia Pueblo Reservation when another car pulled up alongside him at the other side of the pump. Two Native American ladies were on board; one was wrestling with an infant to get his diaper changed. The ladies didn't notice the dinosaur on the back of the pickup, but the infant did. He stopped squirming and his eyes bugged out when he noticed the dinosaur staring down at him. His mother, oblivious to the cause of his distraction, took the opportunity to conduct her motherly duties. Dave's engine was badly out of tune, and when he started up the truck, his muffler issued a cherry bomb explosion that blasted a nasty black cloud of fumes into the open window of the ladies' car. This unfortunate event caused the startled child to pee all over himself and his mother. The driver and the mother, clutching her still peeing and now crying child, ejected out their car like top-gun pilots.

John and I had watched the scene unfold from our car next to them. The mother, laughing and wiping herself off with her free hand, asked us, "What was that?"

Without missing a beat John told her, "From here it looked like a dinosaur had a gas attack."

INDEX

ABOUT THE TYPE

THIS BOOK WAS SET IN WEISS, A TYPEFACE DESIGNED BY A GERMAN ARTIST, EMIL RUDOLF WEISS (1875–1942). THE DESIGNS OF THE ROMAN AND ITALIC WERE COMPLETED IN 1928 AND 1931 RESPECTIVELY. THE WEISS TYPES ARE RICH, WELL-BALANCED, AND EVEN IN COLOR, AND THEY REFLECT THE SUBTLE SKILL OF A FINE CALLIGRAPHER.

ABOUT THE
AUTHORS

LOUIE PSIHOYOS was born in Dubuque, Iowa, on Income Tax Day,
1957. He was hired out of the University of Missouri Journalism
School by *National Geographic* three credits shy of a degree in photojournalism. He lived in New York City for ten years before moving
to the island of Antigua, which served as his base while he traveled
the world working on this book. He now lives in Boulder, Colorado,
with his wife, Victoria, a former dancer with the New York City Ballet, and his two sons, Nico and Sam. His photographs have appeared
in numerous publications and hang in museums and galleries around
the world.

Interested in the road least traveled, and with no scholarly recogniz-
able claims to fame, JOHN KNOEBBER lives at sea level on a sailboat in
Sausalito, California, when not trotting the globe.